T0329747

Understanding Collective Decision Making

Understanding Collective Decision Making

A Fitness Landscape Model Approach

Lasse Gerrits

Department of Political Science, Otto-Friedrich University Bamberg, Germany

Peter Marks

Department of Public Administration, Erasmus University Rotterdam, the Netherlands

Edward Elgar
PUBLISHING

Cheltenham, UK • Northampton, MA, USA

Published by
Edward Elgar Publishing Limited
The Lypiatts
15 Lansdown Road
Cheltenham
Glos GL50 2JA
UK

Edward Elgar Publishing, Inc.
William Pratt House
9 Dewey Court
Northampton
Massachusetts 01060
USA

A catalogue record for this book
is available from the British Library

Library of Congress Control Number: 2017931754

This book is available electronically in the **Elgar**online
Social and Political Science subject collection
DOI 10.4337/9781783473151

ISBN 978 1 78347 314 4 (cased)
ISBN 978 1 78347 315 1 (eBook)

Typeset by Servis Filmsetting Ltd, Stockport, Cheshire

Printed on FSC approved paper

Printed and bound in Great Britain by Marston Book Services Ltd, Oxfordshire

Contents

Acknowledgements

This book actually didn't start as a book. It started as a casual conversation on a lazy afternoon when we tried to do a quick sketch of a fitness landscape for the social sciences on a scratched whiteboard. It didn't work. And so we thought we should try a little more. How hard can it be anyway? That was five years ago. The initial idea resulted in a full research programme that has become our main source of scientific inspiration and joy. This book is our own work, and we take sole responsibility for the contents. However, various people have made invaluable contributions and we would like to use this opportunity to thank them.

First and foremost, we would like to thank our families for their extraordinary patience and endurance, even allowing us to use holidays as an excuse to push the research further. This is more than we deserve, really. A similar kind of patience was also present with our publisher Edward Elgar, in particular with Alex Pettifer, who understood that good research requires a lot of time (which is just a neat way of saying that we missed our deadlines by a mile . . . sorry!). We are proud that our book has become part of the portfolio of such a reputable publisher.

We are very thankful to our supervisors and critical but supportive reviewers Sergey Gavrilets (University of Tennessee) and Geert Teisman (Erasmus University Rotterdam). They represented the two far ends we aimed to unite in this book: fitness landscapes as a model from theoretical biology and a practice-oriented understanding of collective decision making. Without their patient help, the book would have lacked in many places. We hope we have managed to meet their high standards, perhaps that we have even reached the level of 'real science'.

Julian Stieg (Otto-Friedrich University Bamberg) deserves all credit for developing un-code.org. We originally just set out to have an online place to store our raw data, but Julian turned it into a mature data processing and visualization tool, free for everyone to use. In addition, he helped out with translating and sorting the raw data of the Gotthard case. Julian has been a very valuable team member who brought new skills to the project.

We would like to thank Wouter Spekkink (University of Manchester) for his time and creativity during the early phase of the research when many ideas and theories were still shifting shape day by day. Those brainstorms were immensely helpful in focusing the research. Wouter's Event-Sequence Database (ESD) was a major source of inspiration for the way un-code. org works. We also would like to thank Mary-Lee Rhodes (Trinity College Dublin) for productive brainstorms about fitness landscapes and an extraordinary Irish barbecue (Irish, in the sense that it rained, but that didn't make the food any less delicious!), which helped us zoom in on the questions that matter in public administration.

We already mentioned that Julian helped us out with the raw data pertaining to the Gotthard case, which was not an easy task because many policy papers of the local communities were hard to come by. Other people also helped with the empirical studies. Sumet Ongkittikul (Thailand Development and Research Institute) very generously introduced us to key people working with the Airport Railway Link as well as scientists in urban and transport planning in Bangkok. Interviews for the Sports in the City study were done together with Iris Korthagen (Rathenau Institute), which was not only useful but above all fun.

We are thankful for all the help we got and humbled by the interest our research generated. We hope we can live up to the expectations.

Parts of this research were funded by the Netherlands Organisation for Scientific Research, grant no. 451-10-022.

1. An uphill struggle

1.1 KING OF THE HILL . . . FOR A DAY

Our story starts with snow, and lots of it. In fact, there was nothing really special about it when it fell during winter 2013 because it was exactly the same kind of snow that falls every year in Europe during winter. However, for Netherlands Railways (NS), it was disastrous. NS had just proudly introduced its brand-new Fyra high-speed train sets for passenger service, and the snow brutally exposed the train's many weaknesses. It collected in the air vents, tore off the steel casing that was supposed to protect the equipment under the carriages, and played havoc with the electronics. However, the trains would have failed even without snow. Earlier, when the weather was still fair, roof plates had come off during testing, as had one of the access doors. But things were also wrong inside the train. The inner doors separating the compartments did not always open when required, and some lavatories were installed incorrectly. When trains were stowed at the railway yard during the night, batteries underneath the carriages had caught fire. Come spring 2013, NS was forced to admit that it couldn't get the trains back into shape. It appeared that there were too many design and construction flaws. The train's constructor, Italian rolling stock manufacturer AnsaldoBreda, had been offered many opportunities to fix the flaws but never really delivered satisfactorily. Consequently, and years after the original deadline to deliver operational train sets, the contract with AnsaldoBreda was formally terminated in August 2013. This put NS in a situation where it had to run a high-speed railway concession costing about 100 million euros per year with neither the proper trains to do it nor the time to fix the problems. In the end, the Dutch government had to step in to rescue NS from going under completely. This created real financial troubles for both NS and the Ministry of Infrastructure and caused distrust among passengers who were left in the cold.

The Fyra train sets were to be the concluding piece of an ambitious project to build a high-speed rail connection between Brussels and Amsterdam. The Netherlands has a relatively solid reputation when it comes to planning and implementing complex projects such as this one. So how exactly did this problematic situation come about? We need to look into the past for answers. Following the first successes of the Japanese in the 1960s, and later the French and Germans, the Dutch government decided to jump on the bandwagon in the late 1970s and to build its own network of high-speed railways. It was obvious that this was going to cost a great deal of money. One way of dealing with these costs was by deploying financial schemes that were novel to the Dutch situation. An important decision was to tender the concession to operate the network instead of granting it directly to the incumbent operator, NS, as was done traditionally. This decision put NS into a new situation where, for the first time in its history, it had to compete with other market parties for the right to operate train services.

The tender, which the Ministry expected to grant for approximately 100 million euros per year, attracted interest from other operators such as Deutsche Bahn from Germany, Stagecoach from the United Kingdom, and SJ International from Sweden. Under pressure from the Parliament and from the then-popular sentiment that railways in the Netherlands should never be operated by a foreign company, the Minister allowed NS to hand in its bid before the auction. This gave NS the opportunity to grab the concession before its rivals could outbid it. However, the preliminary bid submitted by NS was considerably lower than what the Ministry had in mind. More than just a little annoyed, and publicly scolding NS for being 'arrogant', the Minister rejected the offer and started the auction in earnest.

Now what? NS was suddenly under intense pressure to win the concession and understood that it was not going to win the Minister's sympathy just by being the sole Dutch operator on the playing field. Close to the auction's deadline, and in a bit of a panic, it submitted a new and substantially higher offer that would gain the Dutch state 160 million euros per year. In contrast, the competing offers all floated around 100 million euros per year. Acting quickly, the Ministry accepted this unexpectedly high offer and even persuaded NS to settle at 148 million euros because it sensed that the very high offer could spell financial trouble for NS in the long run. But even that lowered price was still almost 50 million euros higher than the Ministry had expected to extract from the concession, which made it look like a good deal for the government.

As for NS, it had now become king of the hill. It had defeated its rivals, complied with the demands of the Minister and finally got the most coveted right to operate the high-speed railway link, which NS deemed very important for its future operations. Now it was time to deliver. With such an expensive concession, it was obvious that the actual revenue service had to be as efficient as possible. NS looked at buying high-speed train sets from established manufacturers, but the price and operational costs of such trains were deemed too high and wouldn't allow NS to earn back the price of its concession. This was an opportunity for AnsaldoBreda to offer a new design that, on paper, could deliver a high capacity and short travel times whilst remaining below the set prices of other manufacturers. In fact, NS had not much of a choice after two of the competing manufacturers retracted their offers. This drove NS into the arms of the Italians, who set out to build the ill-fated Fyra train. The construction process took years longer than envisaged, partly because AnsaldoBreda had never designed and built such a train before, and partly because the designs were revised during construction. After extended trials on the Czech Velim test track and on the Dutch network, the Fyra was finally accepted for commercial service on 9 December 2012. The first passengers were received with cake and drinks, and there was much media attention. After 20 years of deci-sion making, designing, constructing and calculating, the Dutch finally got their own high-speed railway service. Then the snow came.

1.2   SURVIVING IN A DYNAMIC LANDSCAPE

By some measures, NS had actually been quite successful. After a long struggle in an ambiguous situation, it had outdone its rivals and for a brief moment it was king of the hill. However, the successful end of the struggle also triggered a new situation in which the stakes had changed substan-tially. The former strategy of promising something better than its rivals was no longer relevant. It now had to deliver on its assertion that it could run a viable service. Its former competitors had left the arena, and new adver-saries had emerged. NS no longer had to stay ahead of Deutsche Bahn or other operators, but it had to prove itself to the Ministry and, above all, to its passengers. In short, the execution of the concession meant a reset of the actors involved, their relative positions and what they aimed to achieve.

We can use the analogy of hill-climbing in a mountainous landscape to get a better understanding of the decision-making and interaction process that led to the rise and fall of NS in this particular case. NS and other train operating companies competed for the optimal outcome, namely getting the concession. To them, the highest peak constituted winning the concession, so they set out to find the best route to reach that proverbial summit. Similarly, the Ministry had to make moves in order to reach its own particular peak, that is, get the highest return for the concession so that it could recoup some of the enormous construction costs.

This mountainous landscape turned out to be quite dynamic. Once NS had reached the peak of winning the concession, the landscape changed with the introduction of different aims, stakes, actors and conditions. This metaphorical hill-climb was done; a new one presented itself in the shape of running the concession successfully. To NS, it meant that it had to work hard to reach a new optimum or peak in the changed landscape, that is, to turn the concession into a success. The very same strategy that had made it king of the hill now turned into a liability, as NS was unable to develop a revenue service with which it could fulfil its annual payment to the state whilst delivering reliable services to its passengers.

Students of human behaviour will not be surprised by this. Most of the time, people make decisions that they believe will give them a clear return in the foreseeable future. But exactly how those decisions pan out in the long run is usually obscured by the fog of the future. In addition, the positions of actors in the landscape are mutually dependent; that is, one's own position is conditional on where others are positioned. If those others move, the landscape may move too. In the face of such dynamics, myopic decisions are inherent to human nature. Consider how NS was fully focused on winning the tender because of the pressure exerted on it by others, and only considered the issue of buying the right trains after it had won the concession. In terms of the hill-climbing analogy, it means that the actors try to estimate where the highest peak in the current landscape is positioned so that they can determine how to get there. But what may constitute a high peak in the short run may turn out to be minor peak in a landscape with other higher peaks in the long run, something which could not be seen because people have difficulties predicting the future. As time moves on, the actors struggle to determine which peak they will have to climb using certain routes, understanding that the landscape will shift over time and that peaks will change as a consequence of those shifts. Climbing mountains therefore constitutes an adaptive walk, a changing route through a changing landscape in an attempt to gain the best position relative to others.

Let us also consider the daily experience of the people working for organizations like NS or the Ministry in such a landscape. From start to finish, the whole project had lasted for over 25 years. Very few people were involved continuously during this long period. Throughout these years, the project saw a succession of no fewer than ten Ministers of Transport. Some stayed in office for two full terms; some were forced to step down after less than one year. From their perspective, the project was a lumbering behemoth, a moving train – if you'll excuse the bad pun – they could ride but not really steer. They had to deal with the situation they found upon entering office and left a somewhat altered situation to their successors. The process will have appeared as a slowly unfolding one, with no apparent end state in sight until that end presented itself in a rather undesired fashion. They will have experienced the pressure from their environment to move in one direction or another. They will have seen opportunities and threats, and unexpected dead ends. They will have banked on other actors staying put where they didn't, or the other way around: hoped that actors would move where they decided not to. There was probably a sense of direction towards resolution, but such a resolution could also move further into the future as new obstacles emerged and new quick fixes had to be devised. A solution devised for a problem today could be regarded as an obstacle later on. As the process plodded on, a clear end state seemed evasive.

1.3 ON EVOLUTION AND COLLECTIVE DECISION MAKING

The hill-climbing analogy used here is just that: an analogy without much in terms of explanation. It does offer narrative power, because it can convey the complexity of actors trying to align or divert in changing circumstances in an attempt to reach their goals in a fairly accessible way. It is not hard to imagine the actors in the Dutch high-speed railway case as short-sighted mountaineers who struggle to improve their situation while the landscape evolves slowly too. The analogy also invites all sorts of complementary narratives about the interactions between actors, for instance that there may be different routes to the same peak, or that cooperation can help actors to reach their peak more quickly, or the reverse: that walking alone will provide a shorter route to a higher peak. Informative though it is, it doesn't render much explanatory power. We would therefore like to move beyond mere analogies. Anybody trying to understand a process of collective decision making where winners can be losers, and that features an overall outcome that no one had wished for, will be hard pressed to find a fitting framework with which to render explanatory power for such a complex puzzle.

Intuitively, the process can be understood as an evolutionary process. There is a relationship between the considerations and actions undertaken by the individual actors and the overall progression through time. Most, if not all, actors will also have experienced a disconnection between what they did individually at a given time and place and the long-term development of the project. Many of the dynamics of the project stretched beyond their time horizon and outside of their span of control. Still, things happened, and there was a progression, if not necessarily improvement, through time. There was pressure on the actors to make decisions, and there were multiple possible outcomes at any given point in time, with some more likely than others. Long-term development, punctuated change, and a non-linear relationship between individual actions and the dynamics on the population level: this ticks all the boxes of an evolutionary theory. We follow Sanderson's point of view (1990) that many theories in the social sciences bear the hallmarks of evolutionary thought. Indeed, we believe that collective decision-making processes are examples par excellence of evolutionary processes in the social realm. And by using that term – evolution – we point not at a general understanding of social processes as being long-term and unfolding to some distant point in time but rather specifically at the mechanisms that govern the evolution of collective decision-making processes and the understanding that such processes develop because of the selection pressures exerted on it. These properties can be named and can be used to generate explanatory power with regard to the slow unfolding of such processes into an uncertain future.

If collective decision-making processes are to be understood as evolutionary processes, this raises a whole range of intriguing questions. How do such processes unfold exactly? By what mechanisms are they governed? Do these mechanisms contribute to a directional or a functional law, that is, do they have their own futures locked within or not? What exactly is the relationship between the activities of individual actors and the outcome? What is selection pressure and how is it processed? How does one assign fitness to certain outcomes? Approaching collective decision making from an evolutionary angle will give us a better understanding of the kind of conundrums found in many cases, such as the botched Dutch high-speed railway project, if we are able to dissect the evolutionary mechanisms at work. The aim of this book is to present an evolutionary model of collective decision making, rooted in a naturalistic understanding of empirical cases. To this end, we will deploy models and tools from evolutionary theories. Roughly speaking, such theories come in two variants. The first one is very precise in mapping the exact relationships between actions and outcomes, but suffers from being overly mechanistic and from an overreliance on very simplistic and static assumptions about reality for the models to work. The second one leaves much more room for the provisional and contextual nature of such relationships but suffers from an overemphasis on chance and randomness, and requires constant semantic innovation to suggest that the ordinary is extra-ordinary. We would like to mediate between these two extremes and to offer a third way that has the precision of the first variant without its gross simplifications, and that has the attention to the situated nature of decision making of the second without its suggestion that each action or event is unique.

Naturally, and as we will explain in detail in the next chapter, we are neither the first nor the last to be working on these themes. There is already a venerable body of knowledge on social evolution, and our book will not conclude all the debates. On the contrary, we wish to provide more fuel for those discussions. Our specific contribution to this body of knowledge is that we will deploy one of the versatile models from evolutionary biology, the fitness landscape model, to analyse collective decision making. In evolutionary biology, fitness landscapes are used to study speciation and adaptation, that is, the emergence of biological diversity out of common descent and the occurrence of differentiation. Speciation is governed by a complexity of factors, including but not limited to the interaction between environment and species, the internal genetic composition of the species, and the adaptive capacity of the species in the face of (slowly) changing circumstances. A fitness landscape model provides, simultaneously, a model, tool, heuristic, visualization and metaphor with which to analyse that complexity. If applied well, fitness landscapes can also function as the proverbial Swiss army knife for dissecting the intertwined aspects of decision making.

The caveat for this application is in 'applied well'. Transferring a model from biology to the social sciences requires more than just a few considerations and steps. The original model must be understood first, and then transformed and operationalized to suit a different topic. It must be matched with a research method that does justice to social complexity. Most importantly, it needs to be put to the test. We will present a fitness landscape model for collective decision making that (1) facilitates a structured and systematic analysis of collective decision-making processes and (2) allows for an accessible visualization of such processes. To us, the visualization component is at least as important as the analytical component in this day and age where science has progressed beyond overly simple narratives of how actors make decisions. There is a need, now more than ever, for a method to represent such complex processes in a comprehensible yet accessible way as audiences grapple with increasingly versatile reports of how and why people engage in collective decision making. Here, visualization offers a new avenue to accessibly present investigations to wider audiences. Fitness landscapes hold much potential for both visualization and the analysis of collective decision making. This book aims to unlock that potential.

1.4   OVERVIEW OF THE BOOK

We will develop our argument in a number of steps. It is necessary to have a closer look at the nature of evolutionary theories first, and to assess how such theories can also inform those who would like to study social processes in general, and collective decision making in particular. Also, we will highlight the origin of the fitness landscape model and discuss the ways in which the model is used by others and can be used for our specific aims. This is the core theme of Chapter 2. Chapter 3 covers the philosophy of fitness landscape inquiries and presents the ontological and epistemological foundations of our particular approach. We present the actual model and its details in Chapter 4. In Chapter 5, we demonstrate the main principles of our model, both its basics and its dynamics, in a highly detailed narrative about the attempts to build and operate a high-speed railway in the Netherlands – a case we have already introduced earlier in this chapter. A closer look at three specific dynamic mechanisms of the model is given in Chapter 6, where we highlight each mechanism by demonstrating its value in three empirical studies: local communities in the Gotthard region, Switzerland trying to develop a vision for the future of their region; the city of Rotterdam, the Netherlands trying to realize a sports campus in the city; and the Thai government trying to foster economic growth through the development of an airport, a railway link and an urban district in Bangkok. We synthesize the findings from the individual studies into a characterization of the evolutionary nature of collective decision making and present six archetypes of such processes in Chapter 7.

This book is the result of five years of theoretical and empirical research. Among other undertakings, we have carried out an extensive literature research (Gerrits and Marks, 2014a, 2015), developed a model, and carried out five major empirical studies on the basis of written sources and interviews (Gerrits and Marks, 2014b; Gerrits et al., 2015a, 2015b). We didn't want to clutter the main text with too many details about the sources, so a list of all sources as well as the ways in which we processed the data is supplied in the appendices. As we explain in more detail in Chapter 3, we have decided to work with qualitative data. In order to handle the consequent vast amount of data and to be able to render visuals from that pool of information, we developed an application with which one can structure, code, score and visualize case-based data. This tool is available for all readers to experiment with at www.un-code.org.

Now, let's get to work.

2. Models of social evolution: fitness landscapes

2.1 EVOLUTIONARY THEORIES IN THE SOCIAL SCIENCES

To most people, the term 'evolution' will be closely connected to the work of Charles Darwin, his voyage around the world in the *Beagle*, his observations regarding the variation of species, even his bearded image perhaps. His book *On the Origin of Species by Means of Natural Selection* (1859), in which he proposed common descent and subsequent variation, natural selection and retention in order to explain nature's diversity, laid down the foundations of contemporary evolutionary biology. Unfortunately, many also believe that Darwin's ideas, and evolutionary theories in general, are restricted to the domain of biology. Evolutionary theories would therefore 'not apply' to the social sciences. This divide between biology on the one hand and the social sciences on the other is rather unfortunate, because it means that one forfeits a theoretical framework that holds considerable explanatory power. Also, the divide is a thoroughly artificial one (e.g. Byrne and Callaghan, 2013 for an extended discussion). Allow us to elaborate.

Darwin's theories did not come out of nowhere. Before Darwin, there were others who had explored proto-evolutionary theories, among others his own grandfather Erasmus Darwin and the French biologist Jean-Baptiste Chevalier de Lamarck. In fact, Lamarck is often credited with articulating the first cohesive theoretical framework for evolutionary thinking, among others describing adaptation to local environments and processes of differentiation. As Charles Darwin demonstrated later, Lamarck had been on the right track but mistaken about selection, because he believed that the use or disuse of certain traits would determine selection, while selection is, in fact, blind (Ghiselin, 2009; Hodgson and Knudsen, 2006). However, it does not mean that his general ideas should be discredited. On the contrary, scientists such as Lamarck paved the way for Darwin's theories about evolution.

An important moment in Darwin's thinking came when Captain Fitzroy planned a second surveying tour for map-making and various other observations. He invited Darwin to come along as a naturalist and fellow traveller, and Darwin happily obliged. During the five-year-long voyage, he observed, collected and, importantly, met other people in remote parts of the world. Drawing on Darwin's travelogue *The Voyage of the Beagle*, Ghiselin (2009) identifies a considerable number of such social encounters that contributed to Darwin's ideas about evolution. For example, he noted the major differences between inhabitants and 'civilized people' at Tierra del Fuego, and, upon meeting slave owners in both Brazil and Cape Town, concluded that similar economic circumstances can lead to similar customs despite considerable geographical differences. The voyage brought together the materials from which he would write *On the Origin of Species*. While Darwin's theories concern common descent, speciation and differentiation of species in general, Ghiselin argues that these theories are as much informed by his social observations as by his observations regarding animals and plants:

> Instinct provides the basic faculties that are necessary for civilized man to come into being. Learning and inherited habit are responsible for material culture, including domesticated plants and animals as well as technology, which provides the basis for a better standard of living, but only if there is a certain amount of social control and an end to anarchy is much improvement apt to occur. Customs and practices are adaptive, and reflect of the economic situation. Under the appropriate circumstances people become more civilized, morally developed, and cultivated. (Ghiselin, 2009: 6)

When Darwin was asked to join the world survey, the Victorian era was already in full swing. This was an age of prosperity and scientific progress, and it is safe to assume that observations about this society and the theories of his peers had primed Darwin even before he had set sail on the *Beagle*. Adam Smith, for example, was one such source of influence on Darwin. Smith, a philosopher and political economist, wrote a book in which he attempted to trace the roots from which nations derive their wealth. *The Wealth of Nations*, the shortened but unambiguous title under which the work had become famous, became very popular and influential. Darwin would find inspiration in Smith's idea that the individual's efforts to pursue self-interest may frequently benefit society more than if the person's actions were directly intended to benefit society; that is, self-interest is not at odds with the greater good and could help in attaining group fitness. Smith's theory stresses that the choices and actions of individuals are restricted by a short time horizon and span of control, as myopic or short-sighted self-interest doesn't take into account the group's long-term perspective, yet still relates to the group's likelihood of survival. While Smith was not in search of an evolutionary theory per se, his ideas have a decidedly evolutionary ring to them.

We should also consider the ideas of Herbert Spencer, a philosopher, biologist, anthropologist, sociologist and political theorist. Spencer developed an all-embracing conception of evolution as the progressive development of the physical world, biological organisms, the human mind, and societies. He is known for the notion 'survival of the fittest', a phrase Darwin borrowed, which means that those who fit best in a certain niche will endure. Spencer didn't make a clear distinction between the social and physical world as we are often prone to do nowadays. He rejected the idea that human history is solely marked by unique events and believed that a kind of comparative sociology would highlight that there are recurring themes and fixed stages in history, that is, that history does indeed repeat itself. He posited that all structures in the universe develop from a simple, undifferentiated homogeneity to a complex, differentiated heterogeneity, while being accompanied by a process of greater integration of the differentiated parts (Sanderson, 1990). He identified four stages of differentiation that societies (presumably) would have to go through. These four stages would describe human development from that of rudimentary societies with little organization to that of the great civilized nations. In other words, differentiation was the key concept by which one could understand the development of life at large. He was not the first to suggest that – the basic idea was already present in Lamarck's work – but the notion was central to Spencer's theories, and he believed it was applicable to the biological as well as the social realm. While such theories may now seem a little naïve given the current state of the world, it should be pointed out that they were developed during a specific period in British history when changes in societies could be observed and experienced and when, generally speaking, current generations could be better off than their ancestors.

Such ideas were not limited to British Victorians. For example, Sanderson (1990) points to Lewis Henry Morgan, an American anthropologist and social theorist who worked as a railroad lawyer. His work on kinship and social organization combines the ways in which people organize themselves with the deployment of technology, an argument in which there was ample attention to the emergence and role of governments. He posited that societies evolve through a number of discrete stages, which he called 'ethnical periods': savagery, barbarism and civilization. Note the correspondence to Spencer's attempts to discern discrete stages in the evolutionary trajectories of societies. Morgan made a principal distinction between two types of coordination: *societas*, where kinship is the central organizing principle of life; and *civitas*, where the state has taken over many of the tasks and functions originally assumed by kinship (Sanderson, 1990).

Other names could have been listed here, for example Alfred Russel Wallace, a naturalist, geographer, anthropologist and biologist who can be credited with independently developing a theory of natural selection very similar to Darwin's, and with urging Darwin to publish his results quickly. But the main thrust of our argument will be clear by now: these original thinkers were not very much concerned with the question of whether they were biologists or social scientists, and their evolutionary theories were as much about the development of societies as they were about the development of species. Preliminary ideas about differentiation, variation, selection and retention, not to mention fitness, were already present in those theories. Importantly, the evolutionary framework was not the work of one person but rather the socially constructed product of a large group of thinkers, a framework that could emerge at that particular point in time because all the circumstances aligned favourably. Their ideas were shaped in a particular socio-cultural setting, so it is only natural that those ideas and the terms used to express those ideas reflect certain cultural values.

The moment a new theory is proposed there will be numerous protestations, counter-arguments and evidence to the contrary. Evolutionary theories were no exception. Sanderson (1990) notes a number of important criticisms that are worth revisiting here, because they inform us about what an evolutionary theory in the social sciences should constitute. The most important criticism concerns the difference between directional laws and functional laws in evolution. The idea of directional law holds that societies evolve towards a better state through various (perhaps even fixed) sequences of stages, and because of the potentials inherent to society that were previously unlocked. Functional laws don't feature such an unfolding, and there is no actualization of inherent possibilities. They 'rather attempt to explain historical changes as the result of particular factors operating in particular ways within the context of particular sets of constraints' (Sanderson, 1990: 17). One could argue that some theorists were unclear about whether the evolution of societies concerned a necessary trajectory or whether evolution just denoted long-term change under certain conditions.

This takes us to a second criticism, namely that one could believe that such evolutionary theories envisioned societal progress as a necessary evolutionary outcome, a kind of 'doctrine of progress' as Sanderson (1990: 30) calls it. Indeed, many were ambiguous about this or simply didn't consider it something they needed to discuss. For example, Morgan believed that, on the whole, societies would progress towards a better end-state and that people or governments could undertake certain activities to unlock that potential. Similarly, Spencer believed that conflict and struggle would give rise to fitter and better kinds of society. Naturally, such ideas could only be proven through comparison (do other similar societies display similar stages?), and the focus should therefore shift from the evolutionary outcome towards the process of evolution.

Having said that, it can also be argued that the original thinkers were often poorly understood or reinterpreted, and that they didn't mean to imply that all societies necessarily have to progress from savagery to civilization and that there are (by definition) inferior people. For example, they didn't say that all societies have to develop in the same way, but rather that there are similarities in the evolutionary trajectory of different societies. Furthermore, evolutionary theories have always suffered from confusion concerning the difference between descriptions of such evolutionary trajectories on the one hand and the identification of causality or even general laws that govern such evolution on the other hand. In this respect Sanderson highlights Spencer, who explains that the move from one stage to another stage is not 'necessary' but could take place when certain conditions are met. On top of that, there is actually evidence that for example a great technological breakthrough can propagate itself under certain conditions, helping shape societal dynamics (Bijker, 1997; Geels, 2002).

A third criticism that can be levelled at theories of social evolution is that they encourage eugenic and, worse, racist and elitist worldviews. Eugenics is a controversial area where theoretical ideas about variation, selection and retention merge with normative stances about the promotion or elimination of certain human traits. It is certainly possible to go through some of the classical works and cherry-pick certain quotes that could support such normative stances, but it is equally possible to find quotes pointing to the contrary. We don't aim to settle the score here. Our point is that this criticism is one of the main reasons why theories of social evolution drew considerable condemnation from social scientists. Subsequently, evolution in the social sciences fell from grace.

A renewed appreciation or revival of evolutionary thinking in the social sciences can be identified much later in the works of anthropologists such as Childe, White and Steward. They had in common that 'none of them adhered to a developmentalist or unfolding model of cultural change, and thus all offered explanations of evolutionary transformations that rested on an ordinary causal epistemology' (Sanderson, 1990: 96). In addition, they pointed out not only that evolutionary change constituted an unleashing of potential previously locked into the society itself, but that the environment played an important role, too, in particular with regard to social processes such as diffusion. They thus presented a much more refined view of how evolutionary theories can indeed contribute to a better understanding of societal dynamics, and thereby managed to counter some of the earlier criticism.

From then on, evolutionary theories and principles can be seen to resurface in the social sciences, bearing in mind, of course, Sanderson's distinction between evolutionism in theories about long-term change, and evolutionary theories that attempt to explain change or stasis through the use of evolutionary mechanisms. Even one of the great post-war sociologists, Talcott Parsons, was a proponent of such a use of evolutionary principles. This may be surprising, because he is now mostly remembered for articulating a system's theory of social structures that, while influential, was criticized for being essentially homeostatic and for relying too strongly on a functionalist explanation, the social scientific equivalent of teleology in biology, as a necessary condition for explanation (Gerrits, 2012). However, Sanderson argues that his later work featured many evolutionary ideas, in particular because it focused on functional and structural differentiation, that is, the emergence of increasingly complex forms of roles and functions, which is a core principle of evolutionary theories that can't be explained without taking time into account. In his theory of evolutionary universals, he introduced the idea that social systems may evolve to a state where they are better able to deal with environmental pressures, that is, where they have developed enhanced adaptive capacity (Parsons, 1991). Systems that attain higher adaptive capacity may also continue to do so just because of their previous disposition towards adaptive capacity. Such propositions, and the ones about functional and structural differentiation in societies, strongly resonate with evolutionary theories. Even though many of Parsons's ideas were later amended or even rejected, his influence shouldn't be underestimated, if only because certain other social theories have been developed in response to his.

Within evolutionary thoughts about society, it is only natural that some scientists turned to the phenomenon of collective decision making. Let's for a moment focus our attention on some examples of how evolutionary theories can inform the analysis of collective decision-making processes. Armen Alchian (1950) deployed variation, selection and survival in order to criticize assumptions of perfect foresight, profit maximization and utility maximization as guides for decision making in the competition between firms. He backs away from the then-common focus on profit maximization at the individual level (which would require an impossible certainty with regard to future developments) and instead directs his attention to the operation of economic systems or wholes. This perspective focuses on the reciprocal relationships between system and environment, and the types of economic behaviour that are selected over time. Imitation of routines and products, innovation and positive profits are seen as the economic equivalents of genetic heredity, mutation and natural selection. Alchian concludes that

> the economist may be pushing his luck too far in arguing that actions in response to changes in environment and changes in satisfaction with the existing

state of affairs will converge as a result of adaptation or adoption toward the optimum action that should have been selected, if foresight had been perfect. (Alchian, 1950: 220)

A second well-known example can be found in the work of Richard Nelson and Sidney Winter (1982). Similarly to Alchian, they start their argument by criticizing one of the core assumptions of neoclassical economics. That assumption is that changes in demand or supply are instantly responded to. In reality, however, impeded foresight and imperfect knowledge mean that there can, and will, be delays before corrective action is undertaken. Over a number of iterations, this can generate considerable volatility. The problem is compounded by the fact that a corrective action could very well be misdirected, owing to the aforementioned information problems. Thus the normal state of economic systems is one of continuous reciprocal adjustment. This stands in stark contrast to the common view that such systems are principally in an equilibrium state and that deviations from that state are therefore temporal at best. Evolutionary economics offers credible alternatives to mainstream explanations of the dynamics of economic systems (Nelson, 2006; Nelson and Winter, 1982).

A third example, then, comes from the work of Frank Baumgartner and Bryan Jones studying the role of agenda-setting in shaping and changing policies and associated institutions (Baumgartner and Jones, 1993). Using Eldredge and Gould's (1972) theory of biological punctuated equilibrium, they explain why policies and institutions do not live forever, but instead change with the policy agenda. Stability occurs when there is consensus about combinations of problems and solutions and their relative priorities in the public agenda. Conversely, instability is generated when new policy issues are introduced, and is caused also by the portrayal of these issues and where these portrayals can be promoted (Parsons, 1995). The instability phase in the political and bureaucratic systems allows access to the policy agenda and, consequently, the possibility to change it, as well as the institutions that support it. A new period of stability sets in when new issues, the agenda and the supporting institutions are matched with each other (Gerrits, 2012).

Examples such as the ones above, and there are many more like them, show unambiguously that evolutionary theories can generate explanatory power for research into collective decision making. Coming full circle, then, we should point out the work of John Maynard Smith (Maynard Smith, 1982; Maynard Smith and Price, 1973). Maynard Smith adapted game theoretical methods first used to study economic behaviour (von Neumann and Morgenstern, 1953) to model evolution at the phenotype level. The cornerstone for his approach is the evolutionary stable strategy

(ESS), a refinement of the Nash equilibrium. In his adaptation to biology, an evolutionary stable strategy is defined as the strategy that, when adopted by the full population, prevents a mutant strategy from invading under the influence of natural selection (Maynard Smith, 1982). ESS can be used in biology because, he argues, it focuses on the evolutionary outcomes and ignores questions about whether population members acted rationally or not. The only necessary condition for the models to work is to accept that species are primarily concerned with self-interest by Darwinian fitness, which is a very reasonable assumption. Thus ESS turned out to be useful in both biology and the social sciences (e.g. Axelrod, 1984).

In a nutshell, our argument is now as follows: theories about evolution have proven to generate analytical power for phenomena in the social realm by offering explanatory mechanisms that are empirically empty; that is, they are context-independent. Thus there is no reason to limit such theories to biological phenomena alone. Naturally, one has to account for the vagaries of a specific domain – and we will do that in this book, in particular in the next chapter. The origins of evolutionary theories and the subsequent iterations in the social sciences give ample reason to believe that they provide a solid framework or template with which one can analyse the emergence and development of social processes and structures in general, and collective decision making in particular.

2.2   SEWALL WRIGHT'S ADAPTIVE FIELDS AND ITS VARIANTS

The exposition in the previous section should not give the impression that a unified, monolithic evolutionary framework for the social sciences is just around the corner. In fact, one could argue that such comprehensive attempts tend to grind to a halt owing to the sheer complexity of social reality, for example Parsons's work discussed before, or because they require considerable mental gymnastics on the part of the reader, for example Luhmann's work in response to Parsons (Bednarz, 1984; Luhmann, 1977, 1984, 1995). On the contrary, one can observe considerable epistemic fragmentation in the social sciences. Arguably, this is inherent to scientific progress, and it should be pointed out that biology is similarly fragmented (e.g. Petkov, 2014, 2015). As in the social sciences, there are many contradictions, controversies and debates. If anything, both sciences share a lack of final answers, which is not necessarily because of sloppy science but rather because of the nature of the subject.

Having said that, biology did embark on a merging of geneticist and naturalist theories in what has become known as modern synthesis.

Modern synthesis sparked a renewed interest in evolutionary biology, not dissimilar to the resurgence in the social sciences as discussed before. As is often the case, this was not the work of one person alone, although Dobzhansky's *Genetics and the Origin of Species* (1982) is often mentioned as a key publication. For the purpose of the present study, we will focus extensively on the work of Sewall Wright. Together with Fischer and Huxley, Wright is considered to be one of the main driving forces behind modern synthesis (Ruse, 1996). As someone who had combined practical experience of breeding guinea pigs with theoretical knowledge and mathematics, he articulated a theory in which he combined the occurrence of genetic drift in small populations with environmental factors in order to understand evolution as cumulative changes that 'at each level of organization – gene, chromosome, cell individual, local race – make for genetic homogeneity or genetic heterogeneity of the species . . . The type and rate of evolution in such a system depend on the balance among the evolutionary pressures considered here' (Wright, 1931: 158). He called this 'shifting balance theory'.

An important element of this shifting balance theory is genetic drift. Genetic drift holds that small populations are subject to so-called sampling effects where the selection of certain alleles over generations in small groups is partly determined by chance and certain quirks rather than by natural selection for fitness. Sampling effects are much more visible in small groups, whereas such effects would be levelled out in very large groups (Masel, 2011). In combination with earlier work on population genetics (see for example Bacaër, 2010), this led him to write a paper entitled 'Evolution in Mendelian populations' (Wright, 1931). In this paper, he used mathematics to demonstrate that, 'in small populations, certain genes might move randomly up to total fixation or down to extinction, purely because of "sampling effects". And this change in a gene's proportions would occur even though forces of selection, and mutation, were working in the opposite direction' (Ruse, 1996: 370). In other words, he showed that there is such a thing as non-adaptive drift that could result in new gene combinations when selection is sufficiently slight. In his model, fitness depends on the combination of alleles for that genotype and on the conditions of the field itself, that is, the titular 'balance'. Alleles and genotypes could change in response to evolutionary pressures such as natural selection, mutation and migration (Gavrilets, 2004; McCandlish, 2011).

Wright's (1931) paper is now considered an important step towards modern synthesis. But no matter how good it was, many readers found it difficult to comprehend owing to its heavy reliance on mathematics to complete the model. An invitation to present his ideas at a conference meant that he had to reconsider the mode of presentation or risk running out of time with an audience who would probably get lost in the many mathematical formulations. He therefore decided to discard the formalistic approach altogether and to explain the main properties of this shifting balance theory in words, in terms of fields of gene combinations, and deploy the metaphor of the 'adaptive surface' (Petkov, 2015; Ruse, 1990, 1996). To this end, he developed a number of visual representations that were meant to highlight various possibilities for the adaptive surface and to get the message across unambiguously. These were published in the conference proceedings as 'The roles of mutation, inbreeding, crossbreeding and selection in evolution' (Wright, 1932).

The basics of that model are easily explained. It features a gene or a set of genes that occurs in combination with other genes. Assigning values to each genotype enabled Wright to represent the distribution of adaptive values under a particular set of conditions over the space of genotypes in a two-dimensional field of gene combinations. The location of those gene combinations in the adaptive surface is associated with a degree of adaptedness, that is, biological fitness. In the third dimension, this fitness can be represented by peaks and lack of fitness by valleys, thus constituting a surface plot or, as it became popularly known later, a landscape. In such a (metaphorical) landscape, genes will cluster around peaks because of selection pressure and the fact that gene combinations can't sustain low fitness. The original two-dimensional figure from Wright's (1932) paper is reproduced in Figure 2.1.

Besides this two-dimensional representation of the general adaptive surface, the visualizations included six particular configurations of gene combinations within a field of all possible combinations. These varieties show the occurrence of different mutations given changing circumstances, such as a specialized variety that has adapted to very specific niche conditions and runs the risk of extinction once that niche has seized to exist, and a variety that can remain stable when large enough under uniform selective pressures (see Figure 2.2).

Wright was correct in anticipating that these visualizations would offer greater accessibility than the formalistic mathematics of the 1931 version. Together, they formed a potent theoretical and operational model of speciation. Indeed, many authors built on his works and those visualizations, but there has also been considerable criticism regarding the use of the pictures (cf. Petkov, 2015). Whereas to some they provided a framework with which they could structure their empirical findings at that time, others considered them an incorrect representation of the complex mathematics on which his theory was founded (McCandlish, 2011; Provine, 1986; Ruse, 1990, 1996). In short, the debate about what adaptive landscapes are and whether there is utility in them is still ongoing (Petkov, 2015). A short

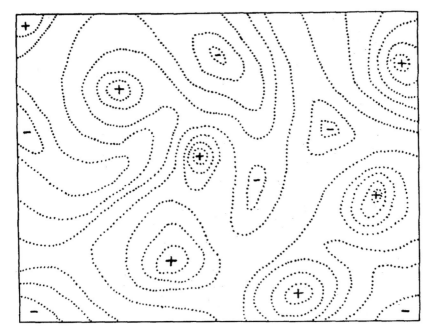

Source: Wright (1932: 358).

Figure 2.1   *The original illustration in Wright (1932) shows a graphical representation of the field of gene combinations in two dimensions. It is a simplified version because, in reality, there are many thousands of possible dimensions. The dotted lines represent contours with respect to adaptiveness*

overview of the most pressing themes is necessary in order to understand our argument in the remainder of this chapter.

First, and foremost, while elegant and versatile, the visualizations can be criticized for being overly simplified, 'not that this should be a matter of any great surprise, given how much information is being crammed together and simplified to get it to work' (Ruse, 1996: 384–385). A two- or even a three-dimensional surface or landscape can only represent a fraction of the immense number of all possible gene combinations (Plutynski, 2008). Besides, while the graphical representation of the landscape looks like a continuous, uninterrupted surface, there are restrictions when trying to work with empirical data (Provine, 1986). While it is possible to map individual data points representing gene–gene–fitness combinations on landscape, the remainder of the surface is inevitably filled up with simulated

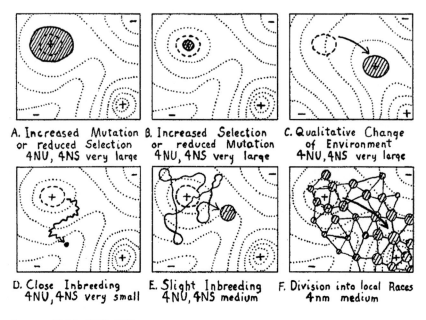

A. Increased Mutation   B. Increased Selection   C. Qualitative Change
   or reduced Selection   or reduced Mutation   of Environment
   4NU, 4NS very large   4NU, 4NS very large   4NU, 4NS very large

D. Close Inbreeding   E. Slight Inbreeding   F. Division into local Races
   4NU, 4NS very small   4NU, 4NS medium   4nm medium

Source: Wright (1932: 361).

*Figure 2.2   Particular types of gene combinations within the general field
               of possible combinations as shown in Wright (1932)*

genotypes. It should be noted here that Wright was well aware of such limitations, and in later work he explored their extent and how they influenced for example conclusions about selection (Wright, 1968). To him, the visualizations were still useful despite not being accurate in displaying the *n*-number of possible combinations. He argued that the low-dimensional representation could serve as an entry point for thinking about the complexity of high-dimensional landscapes that would extend the boundaries of two- or three-dimensional diagrams (McCandlish, 2011).

Second, certain aspects of the adaptive landscape were assigned properties that were amended or even rejected. For example, Wright's original version with its suggestion of high peaks and low valleys would require lateral movements, that is, crossing valleys. Crossing valleys would imply that a temporary loss of fitness is a necessary condition for gaining fitness. This would be possible in small group sizes and through genetic drift, but it would be much harder than originally envisaged. Gavrilets (1997, 2003, 2004, 2010) argued that evolution in such landscapes would rarely occur but would confirm punctuated equilibrium if it took place. As an extension to and specification of the original model, Gavrilets proposed a holey landscape, which can be considered a flattened fitness landscape formed by genotypes with fitness within a narrow fitness band. It is very likely that the actual number of dimensions in the adaptive field is very high, but it is equally likely that there is much redundancy in those dimensions because many of them differ only slightly in ways that don't matter for the outcome. It is therefore not necessary to assign equal weight to each dimension. The holey landscape is then defined

> as an adaptive landscape where relatively infrequent high-fitness genotypes form a continuous set that expands throughout the genotype space . . . The smoothness of the surface in this figure reflects close similarity between fitnesses of the genotypes forming the corresponding nearly-neutral network. The titular 'holes' include both lower fitness genotypes ('valleys' and 'slopes') and very high fitness genotypes (the 'tips' of the adaptive peaks). (Gavrilets, 2003: 148)

One can therefore exclusively focus on viable neighbours. This works well because a higher dimensionality of the landscapes leads to a lower perculation threshold $p > \frac{1}{L}$ where L = the number of dimensions of the landscape as represented by the strings of viable neighbours. This forms the holey landscape that focuses on the bandwidth of mutation where speciation moves around the 'holes'. The holey landscape solves a number of unresolved issues of the archetype model, in particular by presenting an alternative to the issue of 'peak-hopping'. Gavrilets (2004) also argued that an n-dimensional landscape has substantially different properties from a low-dimensional landscape. This questions the assumption that a two- or three-dimensional model can be scaled up to n dimensions unconditionally and without taking into account the qualitative differences between low- and high-dimensional versions.

Third, it should be mentioned that adaptation is a response to past environments rather than an anticipation of the future, and that fitness is not a property of a genotype alone. Adaptation occurs through small steps, which implies that the search for fitness is first and foremost a local search in a rather restricted space of possibilities. This point matters considerably when it comes to the transfer of the model to the social realm, where actors can be assigned reflexive capacities not only to look at the past but also to speculate about future developments.

Fourth, the adaptive landscape metaphor is notorious for the possibility of interpreting and using it in a variety of ways (Petkov, 2015). Wright's narrative moved between fitness landscapes of genetic combinations and of genetic frequencies without acknowledging this, even though these are two different things mathematically speaking (Gavrilets, 2004), and he sometimes moved between statements about the individual level and the group level without being very clear about it (Ruse, 1996). The metaphorical aspect is not necessarily problematic, because it can give a motive and point of entry for modelling and testing, but it can also cause some confusion – a point we will discuss in more detail in section 2.4.1.

Fifth, there is the aspect of the changing of the landscape itself. Wright considered his landscape, if not rock-solid, at least as changing very slowly. Conversely, other authors believe that the landscape can be quite dynamical and that its boundaries are not firm but flexible (Conrad and Ebeling, 1992). Either way, static landscapes may give some basic understanding but are not really interesting for studying evolutionary processes. These, and other such differences of interpretation, show that the adaptive landscape was as clear as it was ambiguous, a point we will be revisiting later in this book.

One could read the points above as failures of the original model. This is not the case. Rather, it shows the scientific process at work here: the original model showed enough potential for scientists to amend, expand and modify it. This may prove Wright's idea that the visualization is as much a heuristic device as it is an actual visualization of his theorem. For example, Dobzhansky (in Plutynski, 2008) used the two-dimensional representation to imagine how, in the distribution of species, each species was located on an adaptive peak separated by gaps of reproductive isolation. Fischer proposed an alternative model where small mutations are much more relevant in the evolution of traits than big mutations, as was commonly thought:

> The survival of a mutant gene . . . is a to a very large extent a matter of chance; only when a large number of individuals have become affected does selection, dependent on its contribution to the fitness of the organism, become of importance. This is so even for dominant mutants; for recessive mutants selection remains very small so long as the mutant form is an inconsiderable fraction of the interbreeding group. (Fischer, R., 1923: 321)

Kimura showed that in neutral evolution, that is, evolution where molecular changes do not influence the fitness of organisms, the number of observed genetic variety per generation is considerably higher (between 100 and 1000 times) than can be expected on the basis of adaptive walks. Therefore 'we must recognize the great importance of random genetic drift due to finite population number in forming the genetic structure of biological populations' (Kimura, 1968: 626). In other words, random walks may be more important in variation than adaptive walks in neutral evolution. Connected to this is the error threshold articulated by Eigen, which expresses a 'rate of error during the reproduction phase below which genetic information is intact and above which it disappears' (Vishnoi,

2013: 59). Genetic codes are very long, but reproduction requires only the first string of genetic information; that is, there is an error threshold. Altenberg (1997a, 1997b) extended the *NK* model into a generalized version which can provide any number of elements and any number of functions. This generalized *NK* model allows the number of fitness components to differ from the number of genes, and allows genes to be added to the genome while keeping the set of fitness components fixed. Mayr (1963) proposed that speciation occurs when a small group of founders move into a new habitat. Central to Mayr's ideas is the thesis that genes not only act but also *inter*act, which is an important addition to the original shifting balance theory.

These examples, out of many, show that Wright's model and visualization have sparked many varieties in which the original theory has been adopted and transformed. In the words of Petkov, the model 'has been the base for plurality of interpretations some of which have overcome the difficulties of Wright's first interpretation, or have been successfully applied to different evolutionary problems' (2014: 2). As befits any scientific theory, some of the ideas were of course falsified by others. However, such findings should not mean that Wright's original ideas and the many amendments to them have been rendered invalid. They were and still are, in Plutynski's words, extremely useful because they served as a template for testing hypotheses and have been central in many attempts to explain and perhaps even predict biological evolution (2008: 620). The question now is how to shift that template from biology to the social sciences. The work of Stewart Kauffman turned out to be instrumental in this shift.

2.3 STEWART KAUFFMAN'S FITNESS LANDSCAPES AND *NK* MODELS

Kauffman and Levin (1987) took into considerations the criticism on the hill-climbing analogy that had been used in Wright's model, to develop a general theory of adaptive 'walks' via fitter combinations. In this version of the adaptive landscape – which from now on we will call 'fitness landscape' in order to keep consistency with the nomenclature – the fitness is not a property of the genotype alone, but also depends on the environmental context. Each genotype is surrounded by a number of other genotypes. 'Adaptive walks proceed from an initial entry, via fitter neighbours, to locally or globally optimal entities that are fitter than their neighbours' (Kauffman and Levin, 1987: 11). Note that there is no genetic drift in this model. In its most basic form the fitness of the genotype is just the sum of the N independent fitness contributions divided by N (Kauffman,

1993: 41). However, in a system with N genes most often the fitness contribution of one gene depends on the other $N-1$ genes. This is the so-called NK mechanism within fitness landscapes in a nutshell. The fitness landscape is rugged when N and K are large, while it is smooth with only one peak when K is zero.

Using this model, Kauffman and Levin (1987) demonstrated the number of local optima, the distance of the adaptive walk to a local optimum, and the alternative optima accessible to entities in uncorrelated landscapes. They also assumed that in many cases landscapes are correlated, that is, that adjacent but different entities have similar fitness. In such instances, fitness is not a property of a genotype alone but depends upon N and K as independent parameters and the environmental context – an idea that they expanded upon. Their findings imply 'that complex biological systems, such as genetic regulatory systems, are "close" to the mean properties of the ensemble of genomic regulatory systems explored by evolution' and that, 'with increasing complexity and a fixed mutation rate, selection often becomes unable to pull an adapting population to those local optima to which connected adaptive walks via fitter variants exist' (1987: 11). In a fashion similar to for example Maynard Smith, Kauffman and Levin also use non-biological examples and methods to demonstrate the dynamics of their model such as the spin glass and the travelling salesman problem. In addition, they use the broader term 'entities' instead of the more strictly defined term 'genotype'. This may invoke all sorts of questions of what type or class of theories they are dealing with and, as they admit in the text, they believe that they are en route to an entirely new kind of theory (1987: 19).

Kauffman and Johnsen (1991) built on the uncorrelated (NK) fitness landscape model to show that the fitness of a genotype is affected by the genotypes of the species with which it is coupled, that is, that the adaptive moves of one agent deform the landscapes of its neighbouring agents. In other words, Kauffman and Johnsen theoretically demonstrate the occurrence of coevolution (Ehrlich and Raven, 1964) between landscapes through adaptive moves on those landscapes. They then aimed to develop a class of models with which this coevolution can be understood, in particular the conditions under which a Nash equilibrium will be established, that is, the situation in which there can be no further advantage in making another adaptive move for any of the genes in the landscape (Maynard Smith, 1982).

The theoretical models described here, together with the ones developed with others (e.g. Kauffman and Weinberger, 1989), culminated in Kauffman's book *The Origins of Order* (Kauffman, 1993).The book's main theme is a search for an explanation of the origins of life and subsequent adaptation and differentiation. To this end, Kauffman deploys a wide range of tools, models and concepts but with self-organization and coevolution as central concepts. Biological order, he argues, is governed by the laws of self-organization, that is, the emergence of structure through mutual interaction between genes without a superimposed design. Consequently, he attempts to flesh out the precise conditional dynamics under which self-organization appears.

The book can be seen as an attempt to supplement – if not replace – the common notion that speciation is exclusively driven by selection. Kauffman's view hinges on the idea of adjacent possibilities. This is best imagined as an attractor basin that contains all possible system states but not the actual system state. These system states are just one further step away from the actual state. 'Once a new state has been achieved in the system by realizing one member of the current adjacent possible, a new adjacent possible, accessible from the expanded actual that now includes the additional member, becomes available' (Kiblinger, 2007: 196). In other words, there exists a theoretically infinite space of possibilities, but the unlocking of those states is conditional and limited to the *adjacent* possibilities. There is thus infinite potential but a more limited actualization of combinations. Novelty emerges from new and unforeseen combinations, which leads Kauffman to claim that such novelty (e.g. in structures or processes) is truly self-organizing. Proposals to supplement or even replace the mechanisms of variation and selection in theories about evolutionary biology are not new as such (Weber, 1998; Weber and Depew, 1996), but Kauffman attempts to model self-organization and selection in such a way that one can investigate how self-organization can enable or restrict natural selection.

Fitness landscapes play a central role in this modelling attempt. Indeed, they may be considered the 'conceptual glue' (Weber, 1998: 135) that keeps the many arguments together. These fitness landscapes are principally not very different from Wright's adaptive landscapes, though Kauffman uses a number of different versions (cf. Kauffman, 1993: 37). Essentially, they are graph-theoretical representations of the reproduction success of genotypes or phenotypes as the physical expression of the genotype. Fitness landscapes model the fitness or replication rate of particular genotypes or phenotypes under selective pressures. In the first case, individual fitness is plotted against individual genotype, that is, the fitness or replication rate of particular genotypes. In the latter case, the landscape is a graph of population mean fitness against the state of the population as measured by allele frequency or trait means (Barton, 2005). The distance between genotypes or phenotypes, that is, the extent to which they are similar or not, and their interactions define the rate of fitness in the landscape. The fitness contribution of each N genes in a genotype depends on the interaction with K other genes and is visualized as a hypercube. Thus fitness landscape models allow researchers to investigate the relationship between diversity, interaction and fitness of genotypes or phenotypes in their environment.

Since Wright, visualization has been an important part of research using fitness landscape models. A popular version visualizes the models as three-dimensional surface plots or a landscape with the population represented on the x-axis and interaction between the genes in the population on the y-axis. Fitness is then represented on the z-axis in the landscape. Each configuration of NK values in the landscape defines a possible individual fitness value. These are assigned either randomly or manually, or the value might be some function of the values taken from each dimension (Calcott, 2008). The fitness values assigned to each combination of values along the dimensions of variation mark out the (multidimensional) surface of the fitness landscape (Calcott, 2008). Organisms move across the landscape in an attempt to gain a better fit by grouping and matching characteristics. However, adaptive moves by one species deform the landscapes of its part-ners (Kauffman, 1993). The adaptation of one organism will influence the success of strategies adopted by other organisms (Haslett et al., 2000). The peaks in the fitness landscape represent the pay-off for optimized adaptive behaviour.

In many ways, *The Origins of Order* (Kauffman, 1993) is a curious book. For starters, it features over 700 pages of dense and sometimes inacces-sible prose and mathematics. It has a strongly self-referential but complex structure (Dover, 1993), which leads some to recommend reading the book non-linearly (Weber, 1998). The sheer number of ideas, propositions and models allows multiple points of access to the text, in turn enabling readers to select specific themes of interest without having to read it cover to cover. While some parts of the book feature tried and tested ideas, the text is also for a considerable part highly speculative, and sometimes confusing and inexact (Alberch, 1994; Weber, 1998). It invites biologists to look at evolu-tion in a particular way, but it provides neither a grand theory of evolution nor a coherent set of proven causal chunks.

Altogether, *The Origins of Order* received a mixed response. The main reason for discussing Kauffman's work so extensively in this book is that the main themes and models have proven to appeal to an audience outside of evolutionary biology. Judging by the many citations and discussions of his idea, the book can be considered the prime means – for better or worse – through which fitness landscapes ended up in the social sciences (Gerrits and Marks, 2014a). This is exactly what Kauffman intended. His ways of phrasing things in sometimes generic, even poetic, ways ('Like the Alps, our landscape here possesses a kind of Massif Central, or high region, of genotype space where all the good optima are located' (Kauffman, 1993: 61–62)) can give readers the impression that his work stretches beyond the borders of biology. In fact, as we have seen in Kauffman and Levin (1987), he was already on track to develop a new class of overarching theories with which the origins of life should be explained. Ultimately, the *NK* model is considered applicable to many types of questions just because it 'allows for a very general description of *any* system consisting of *N* components with *K* interactions between the components and in which there can be any number of states for each *N*' (Weber, 1998: 135, italics in original). In Kauffman and Macready (1995), he used the *NK* model as a sensitizing concept to show that technologies evolve in fitness landscapes that are rugged with conflicting constraints, concluding that there is 'at least an analogy between the unrolling panorama of interacting, coevolving species . . . and the way that technological evolution drives the emergence and extinction of technologies, goods, and services. This analogy can offer intriguing and fruitful insights into the ways that products, organizations, and economies develop' (Kauffman, 1995: 129). He even suspected that these products, organizations and economies are governed by the same or similar fundamental laws. In his view, the *NK* model should be considered as meta-theoretical. Naturally, such a broad theoretical framework would have to explain everything (i.e. life), and separating theories into the isolated silos of scientific domains wouldn't work here – or at least that would be the argument. Kauffman's work may be understood from, and applied to, different domains. Some authors even commented that Kauffman's models and methods are not necessarily relevant to biology (Weber, 1998).

Even though the book may be considered more of an exploratory exercise than a collection of proven theories and causal relations, we need to critically examine his ideas in the light of the theory transfer to the social sciences. Causality as observed and mapped in physics is not necessarily the same as the causality governing the emergence of biological order, let alone causality that drives social interaction. In a similar vein, one may question the assumptions underlying his modelling efforts that rely on models and methods from all kinds of sciences. In addition, empirical proofs are lacking in many areas of his framework, as much of it relies on computational modelling and simulation. Some even doubt whether his propositions could be tested at all (Fox, 1993). Gavrilets's counter-argument that evolutionary processes span such enormous time-scales that they often defy empirical observations and that mathematical models provide an excellent alternative (Gavrilets, 2004) still stands, of course. Still, the proof of the pudding is in the eating.

We want to single out one criticism that is particularly relevant to our book. Through his thematic choices, the phrasing and the use of mathematical models, Kauffman not only implies a general applicability but also risks implying that answers that are generated *from within* his framework must be universally true. Thus it can be that he moves from a theoretical model to claims about the real world. But a mathematical model is just that: a model, a reflection of the intellectual process and an articulation thereof. To be clear, that is a genuinely useful step in the production of knowledge. The idea of fitness landscapes can serve as an entry point into modelling, which in turn allows the researcher to develop and test certain hypotheses. However, it should be pointed out that the soundness of a scientific discourse can't be legitimated through processes and models *internal* to that discourse. Thus Kauffman's attempt at a new class of theories cannot generate its own legitimacy but needs metanarratives from within science in general and even from outside science in the shape of socio-culturally dominant ways of thinking to establish its status (Robertson, 2004). Kauffman's ideas, like others before him, are influenced by the intellectual discourse at this particular point in history and are therefore not self-evident. Shifts in that intellectual discourse may lead scientists to reconsider his claims. Science is not value-free, and empirical and theoretical claims are therefore relative instead of absolute. We need to keep that in mind when discussing his work and the many derivatives presented later in this chapter.

Kauffman's attempt to cross disciplines is not the first, of course. We should point out a similarity to Wright, who also believed that his ideas and theories were applicable to socio-cultural evolution – although he was less explicit about this than Kauffman. Wright made several attempts at building a kind of stratified or hierarchical view of nature in which he tried to classify all sciences by unit of organization with regard to (a) equilibrium and (b) changes within the equilibrium. Wright also claimed to be inspired by philosophers outside of biology, in particular by Bergson and Whitehead (Ruse, 1996). Indeed, Whitehead's process philosophy, that is, the thesis that reality is made up of connected processes rather than of connected material elements, could match with the idea of unfolding evolution (Abbott, 2001). In short, Kauffman's attempts to stretch beyond biology are not alien to that field. *The Origins of Order*, then, forms the key to the theory transfer of fitness landscapes and the *NK* model to the social sciences.

A short summary is in order by now. We started this chapter with the argument that evolutionary theories concern the social as much as the biological realm. We demonstrated how some of the main thinkers in biology were inspired by ideas from the social sciences, and how ideas from biology inspired social scientists. Consequently, we argued that there is no reason to restrict evolutionary theories to research about long-term biological processes. A number of post-war authors in the social sciences showed why and how such theories also work in the analysis of social processes and structures. Naturally, this doesn't mean that there is such a thing as a monolithic evolutionary framework in the social sciences; the same holds true for biology. However, modern synthesis provided an important thrust to integrate various strands in biology. Wright's work was pivotal in that synthesis, in particular his theorem about adaptive surfaces. Kauffman extended the adaptive field into his own particular versions of fitness land-scapes. His models were developed and presented in such a way that they provided a bridge between biology and the social sciences – as such fitting in a long tradition of theory transfer between the two disciplines.

We will now turn our attention to the utility and dispersion of fitness landscapes and *NK* models in the social sciences in order to determine how this can inform social inquiry.

2.4 FITNESS LANDSCAPES AND *NK* MODELS IN THE SOCIAL SCIENCES

A fitness landscape is appealing and versatile in its applications because it features 'graph-theoretical structure and bases in non-integral space, presupposing the connections between the elements in the system under scrutiny to have as much explanatory power in the analysis of its dynam-ics as the characteristics of those elements themselves' (Hovhannisian, 2004: 2). Since his 1993 book, Kauffman's work has gained ground inside and outside biology. The model became adopted most prominently in the realm of (business) economics and the organization and manage-ment sciences (Westhoff et al., 1996). There have also been applications in sociology, anthropology, law, public administration and political sci-ences. Although it is still subject to debate, as discussed above, Kauffman's implication that the framework is universal has been convincing enough for quite a few scholars in the social sciences to adopt it in their research. They start by asking questions such as 'What kind of interaction between humans promotes fit?' and 'How can the alignment between actors such as nation states be explained?' To most authors, Kauffman's work serves as the starting point for answering such questions.

While we found that both Wright and Kauffman took central positions in such types of social scientific research, *The Origins of Order* turned out to be the most popular foundation for most authors (Gerrits and Marks, 2014a). In addition, there are specific authors who can be regarded as key movers instigating a particular interpretation of fitness landscapes in certain domains. These are Auerswald et al. (2000) in economics, Levinthal

(1997) in organization and management sciences, Ruhl (1996a) in law, and Lansing and Kremer (1993) in anthropology. These authors occupy a central position within their respective niches. Some authors have built up a considerable repertoire and solid body of knowledge around fitness landscapes (e.g. Levinthal, Rivkin and Siggelkow), whereas many others have restricted their inquiries to a very limited number of publications. We noted that the use of fitness landscapes and the *NK* model in investigations in the social realm could be rather confusing. There appeared to be very little consistency among others within domains (let alone across domains) as to how the models were applied to certain questions or how the visualizations were used and understood. The metaphor has storytelling capacity and some authors would take such narratives to extremes. The versatility of the model is clear, but exactly how fitness landscapes contribute to a better understanding of social phenomena isn't.

Many of the inconsistencies can be traced back to the fact that authors interpret and use fitness landscapes in distinctively different ways. The flexibility of the model means that multiple points of entry are possible. One could start from the *NK* model and build on its mathematics, or start with the metaphor to develop narratives about hill-climbing, or start with the visualization and develop elaborate depictions of particular situations. In addition, there are multiple possible focal points, for example the relationship between exploration and exploitation, between centralization and decentralization, or between cooperative and non-cooperative behaviour. Naturally, different authors choose a particular point of entry and focus in order to answer their research questions, which gives rise to an enormous diversity of applications.

This wild diversity is not exclusively a characteristic of the social sciences. On the contrary, it also applies to the use of fitness landscapes in biology, as so clearly explained by Petkov (2015). Thus we can tentatively control for domain variance and conclude that it is the model itself, and Kauffman's work in particular, that gives rise to these many interpretations in both the social sciences and biology. Some authors (e.g. Kaplan, 2008) lament this diversity and argue that a proper biological theory shouldn't feature contradictory explanations. We beg to differ. It is only natural that this would happen. Besides, scientific heterogeneity is actually rather useful, and we will advance a more detailed argument for this in the next chapter. For now, we note that the authors reviewed found much utility in using fitness landscapes in their inquiries. Across the domains, we categorized the various interpretations and uses in five main modes of social scientific inquiry. They are: metaphors, sense-making, simulation and modelling, theorizing, and case mapping (also note that, independently from our study, Petkov developed a similar categorization in biology). We grouped simulation and modelling because the distinction between the two matters but not as much as to warrant distinct categories. All modes are present in the domains mentioned above, but particular modes are more common in certain domains than in others. We will discuss them below in order to fully appreciate the depth and versatility of the model.

2.4.1 Metaphors

The metaphorical use of fitness landscapes has been present since the beginning. After all, it was Wright himself who in his 1932 article proposed adaptive fields as a metaphor in an attempt to demonstrate multiple aspects of his shifting balance theory and the mathematics founding that model in an accessible fashion. Metaphorical uses in the social sciences are intimately connected to the three-dimensional visualizations with peaks and valleys resembling a mountainous landscape, where a higher position on the z-axis equals improved fit. Such visualizations invite all sorts of analogies, such as the hill-climbing analogy or the moving peaks analogy mentioned before, both of which are as useful in the social sciences as in biology. Typically, authors engage in narratives that involve stories about how human actors try to climb the peaks in the landscape or how they find themselves in a valley, which is equated to a suboptimal situation from which there is only a steep way out. In such accounts, data often comes from case-based, circumstantial observations, but the metaphor itself is the method.

Examples of metaphorical use can be found in for example Geyer and Pickering (2011), Klijn (2008) and Pascale (1999). Geyer and Pickering draw parallels between grazing herds in search of nutritious grasses as a way to survive, and human complexity in the field of international relations: 'To begin, imagine a landscape that is full of flatlands, valleys and mountains and stretches endlessly into the future. Now, imagine that the valleys represent zones of poor performance, the mountains are zones of good performance and the flatlands are areas of neutral performance' (2011: 13). Such landscapes are populated by actors that actively walk around the landscape in search of improvement. In addition, the landscape is imagined as a conveyor belt that keeps moving from the future towards the present in which the actors are located. A similar metaphor is present in Geyer and Rihani (2010). In a similar vein, and borrowing from Pascale (1999), Klijn (2008) regards public management as an act of 'riding the fitness landscape' (2008: 314), where the task of the manager is to be aware of the opportunities in that landscape and to seize those opportunities in order to let his or her policy proposals materialize.

The power of the metaphor lies in such storytelling. The image of actors walking across hilly landscapes can be understood intuitively and is therefore rather efficient in communication without having to resort to complex mathematics. Thus this image may provide the easiest point of entry for audiences not familiar with the idea. However, metaphors come at a price: their inexactitude. There are no real hills to climb in social organization, and managers are no mountaineers (well, there may be managers who like to climb mountains as a hobby – we can't deny that possibility!). In such accounts, the inexactitude is reflected in the vagueness about the various aspects of fitness landscapes. In most cases, neither fitness nor the other dimensions are defined. The same goes for the often ill-defined agents, a vagueness that can perhaps be traced back to the fact that many things in the real world are essentially information-processing units and therefore agents in the broadest sense of the word.

As Wright already understood intuitively, metaphors serve as a means of theory transfer rather than having causal explanatory power in themselves (Chettiparamb, 2006; Lakoff and Johnson, 2003). However, the accounts we have read are rather careless about the mapping of the source and target domain, and the alterations that take place during transfer between the two. If Kauffman writes that agents group to attain a better fit under certain conditions, he doesn't mean that human decision makers such as managers work together to create a better fit. Such errors are most prominent with writers who derive normative statements from fitness landscapes, for example that people should work together to create a better fit.

2.4.2   Sense-Making

Fitness landscapes are also used to make sense of the messy social reality often both observed by researchers and encountered by decision makers. Such accounts appear somewhat similar to the metaphorical uses but are noticeably different in that they don't draw parallels or analogies but rather try to enhance understanding about a particular case by using the vocabulary of fitness landscapes. In other words, researchers may use elements from fitness landscapes to make sense of data in ways they might not be able to see otherwise; for example, thinking in terms of peaks and valleys might help them to understand why a certain agent was unable to move despite being in an unfavourable position.

This kind of inquiry is used often, and examples can be found across all domains, for instance as an illustration of how managers can deal with conflicts between organization and environment (Samoilenko, 2008), how firms search for strategies (Beinhocker, 1999; Girard and Stark, 2003; Merry, 1999), how students try to develop different strategies and conversation topics in an experimental interaction setting with varying degrees of success (Kindt et al., 1999), or how religious and cultural diversity throughout human history can be understood as individual minds mutually interacting within converging fitness landscapes in an open-ended time horizon (Atran and Norenzayan, 2004). Consider the work of Urry (2008), who uses the language of the complexity sciences to explore the space of future directions for sociology. Fitness landscapes help him to make sense of the impact the automobile has in societies, in particular how it has moved from being simply a means of transport to a focal point of society in terms of for example wealth, convenience and environmental impact:

> Automobility changes the environment or fitness landscape for all other existing and future systems; and it achieves this through its superior capacities to adapt and evolve by comparison with all other mobility-systems, especially those that are much more organized through hierarchical relations . . . [I]t draws in many aspects of its environment which are then reconstituted as components of its system, it is central to and locked in with the leading economic sectors and social patterns of twentieth century capitalism. (2008: 265)

As with metaphors, accounts focusing on sense-making are not very focused on the methodology and processing of empirical data. Data comes from examples, most often in the shape of case narratives and circumstantial observations. Exactly how those narratives or case studies were produced is often not explained, but sense-making seems to create a higher level of detail than metaphors. Consequently, the presented case studies are richly detailed, yet at the same time it sometimes is unclear why certain case-specific aspects were chosen over others in becoming connected to the various concepts from fitness landscapes. Obviously, sense-making is relativistic. The rendered stories present only one version of reality, which suggests looking at reality in an alternative, possibly accurate way (van Hulst, 2008). By presenting this version, the author invites others to render alternative approaches to understanding a certain phenomenon. Existing approaches become altered when observers try to fit in new ideas and rework the approach they adopted. An example of such a dialogue can be found in Babcock (1996), who examines Ruhl's (1996a) argument that fitness landscapes help explain how changes in society are reciprocally connected to changes in law.

2.4.3 Modelling and Simulations

Modelling and simulations constitute the most common adaptation to the social sciences by far. They tap into an aspect of fitness landscapes entirely different from metaphorical use and sense-making. Whereas the latter take their primary cues from the images or visualizations of rugged landscapes and their associated narratives, the former take their cues from the mathematics at the roots of fitness landscapes. The generalized *NK* model, which is of little use in metaphorical accounts or for sense-making, is functional here in particular because it is not constrained by the limits of three-dimensional landscapes that are necessary to generate a comprehensible visualization. Modelling fitness landscapes requires the researcher to operationalize the core aspects of such landscapes, including *NK*, fitness, and the *n*-number of dimensions represented in the model. Contrary to work that focuses on metaphors and sense-making, modelling efforts are very precise and explicit about what constitutes the elements of the landscape. Simulations are a further development of modelling, because they put the model to the test. Typically, researchers do multiple test runs. The results are then subjected to a regression analysis in order to determine which approach, for example search strategy, composition versus decomposition, structure, diversification and so forth, renders the highest fitness in the landscape.

Developing such models helps in exploring the relationships between *N* and *K* in different situations. We found many different applications, such as the study of the properties of different fitness landscapes themselves through understanding decomposition and connectivity, how search tasks should be structured (Chang and Harrington, 2000; Garrido, 2004; Rivkin and Siggelkow, 2002; Siggelkow and Levinthal, 2003) or how search algorithms perform in particular fitness landscapes (Haslett and Osborne, 2003; Haslett et al., 2000), exploring the relationship between exploration and exploitation (Becker et al., 2008; Bocanet and Ponsiglione, 2012; Knudsen and Stieglitz, 2007; Lee and Van den Steen, 2010), and the evaluation of different organization structures (Barr and Hanaki, 2008; Ethiraj and Levinthal, 2004; Lovejoy and Sinha, 2010). However, for each instance a tailor-made model needs to be defined, as the generalized *NK* model is too broad a platform to act as a direct template for any research into social issues.

As may be guessed from the summary above, most examples of modelling come from the realm of economics and from organization and management sciences. However, there are also modelling attempts in other domains. For example, Lansing and Kremer (1993) run a simulation that is structured on the *NK* model (albeit implicitly) with which they map the physical evolution of the irrigation network of Bali. This simulation is functional in developing an explanation of how the irrigation network could emerge without any clear design or superimposed structure – emergence in the truest sense of the word. And Fellman (2010, 2011) used the model by Ghemawat and Levinthal (2008) to demonstrate how optimal decision making in terrorist networks can be countered by dividing the network into separate compartments.

By definition, modelling and simulation efforts are very precise and clear cut about the properties of the landscapes. This forces the researcher to move away from the ambiguities of metaphors and sense-making and helps in gaining a better understanding of the mechanisms of the landscape. However, the very limited number of attempts in which external validity is sought suggests that the leap from model to the real world is rather big. Here, the formal elegance of mathematics is at odds with the messy reality of the social realm. A few examples notwithstanding (e.g. Axelrod and Bennett, 1993; Cartier, 2004), models are usually not based on real-world data, nor are the outcomes validated against real cases. Axelrod and Bennett (1993) model nations as myopic actors in order to understand alignment or divergence. Using secondary data, they demonstrate the alignment of 17 European nations during the Second World War. However, in most other cases numerical values are assigned by the researcher and, while this has demonstrative qualities, there is little reason to assume that such values resemble the real world. This illustrates the difficulty of moving from an operationalized model to conclusions concerning social reality, something that some authors are aware of (e.g. Lansing, 2003) but others don't mention.

2.4.4   Theorizing

Whereas the previous modes of inquiry draw from narratives and operational mathematical models respectively, there is also a group of authors who theorize about fitness landscapes by considering them as whole theoretical constructs to be applied to different situations. They present an alternative approach or framework to particular persistent problems in their domains. With this, they hope to develop new avenues of thought in understanding such problems. Authors who engage in theorizing usually reproduce Kauffman (1993) faithfully, because it offers ample scope for such theorizing, as was always intended by Kauffman himself. Take, for example, the term 'agent'. It can be considered a synonym for 'gene' but can also be understood in a broader sense as any unit that processes information, such as a human or collective of humans. While such diverging interpretations of the same term can cause confusion – recall our discussion about metaphors above – it can also lead to the establishment of a novel theoretical framework in the target domain if the operationalization is done well. In the process, authors engage in a theory-driven debate where they revisit Kauffman's work and amend or alter it to suit their own purposes (Room, 2011, 2016).

Examples include theorizing about survival in competitive environments (Dervitsiotis, 2004, 2007), collaboration in manufacturing chains

(Dekkers, 2009), or the design of work within and between organizations (Sinha and Van de Ven, 2005). For example, Whitt (2009), following the financial and economic crises, proposes a theoretical framework to research adaptive policy making for dealing with volatile markets. Fitness landscapes are then used as conceptual tools to develop analogies between ecologies and market mechanisms, where fitness is provided by market selection. Some accounts are very elaborate. The work of Ruhl serves as an example (Ruhl, 1996a, 1996b, 1997, 1999; Ruhl and Ruhl, 1997). To him, the framework is essential in explaining how law changes over time in coevolution with society. Central to his argument are the gain and loss of fit between laws and society; that is, laws tell people what to do and how to behave, but at the same time those laws emerge from socially accepted norms. The occurrence of shifts in society and laws can mean that fit is attained or lost, in particular because laws change with considerable delays. Fitness is then measured by the extent to which laws achieve their goals. Fitness landscapes are used in many ways: as a way of explaining legal change but also in a somewhat normative way, to show that theorists arguing for stability in legal systems are arguing for something impossible. His exposition doesn't feature any actual measurements, but he goes beyond treating the framework as a mere metaphor or as sense-making.

Theorizing is useful because it offers an attempt at theory transfer between theoretical biology and a target domain. It helps researchers to assess if and how fitness landscapes can inform a particular understanding in their own domain. In their application to social reality, researchers may be less formalistic than those who model, yet their interpretations are less prone to vagueness and speculation. Consider for example Beinhocker (2006), who convincingly theorizes how fitness landscapes help in understanding economics from a complexity-informed perspective. He does not model, but shows a thorough and in-depth understanding of the sources, which makes for a sound foundation of his argumentation. Naturally, as with any such attempts at building theoretical frameworks, empirical proofs have yet to be delivered.

2.4.5   Case Mapping

The number of actual empirical and qualitative case studies is very small, even though fitness landscapes offer ample room for the mapping of cases. Fitness landscapes have dimensions that could be addressed by the rich and detailed data often generated in qualitative case studies – an argument we will further develop in the next chapter. Theoretically speaking, case studies enable the researcher to make fitness landscapes relevant for the

social and behavioural sciences. Real-world case studies render explanatory power and demonstrate the various aspects of such landscapes. But,
while the number of authors doing full case studies is very small in comparison to the number carrying out the other applications discussed above,
they form a distinct category present within each domain. Their accounts
centre on attempts to structure an actual case study in terms of fitness
landscapes.

A good example can be found in Rhodes (2008). She investigated six
cases of urban regeneration in Northern Ireland and the Republic of
Ireland. She attempts to analyse the increasing complexity in urban regeneration, in particular the emergence of new rules and schemes. To this end,
she deploys a sub-variety of the fitness landscape, namely the performance
landscape as first developed and introduced by Siggelkow and Levinthal
(2003). In these performance landscapes, agents pursue their agendas,
interact and achieve different outcomes. Using a standard approach to
qualitative case studies, the results were developed in consultation with
the stakeholders as a way of ensuring reliability. Interestingly, Rhodes's
empirical findings suggest that the relationship between strategic choices
and performance outcomes is not as tightly coupled as Siggelkow and
Levinthal (2003) proposed. The coupling varied over time because of the
efforts of certain actors. In addition, Rhodes found that the boundaries
between system and environment are less clear cut than in for example
Kauffman (1993), which shouldn't come as a surprise for social scientists
who have experienced the fluidity of social boundaries.

The limited number of authors theorizing and mapping cases shows
that, because of its wickedness, properly transferring a complete theory
for theory development in the target domain, and operationalizing it to
map it on to empirical data, is still in its infancy. There are a number of
reasons for that. First, using a fitness landscape requires a leap from the
formal mathematics of the model to the (often) qualitative data from
such cases; that is, it is hard to define what variables should be taken into
account, how they should be operationalized and how the formal model
can address data. Second, the amount of data required is substantial and
needs to be longitudinal in order to tell anything about adaptive walks
and changing landscapes. Third, social reality is inherently messier than
the elegant models suggest, something Rhodes discovered, as mentioned
above. Fourth, assigning a particular meaning to certain events or processes remains at the author's discretion and is therefore easier to contest
than for a mathematical model.

2.5   SCIENTIFIC HETEROGENEITY IN FITNESS LANDSCAPE RESEARCH

The diversity of interpretations and applications demonstrates the concept's enormous flexibility: fitness landscapes are used in many ways, ranging from computational models to argumentative vehicles, and applied to countless subjects, ranging from discussions of Talcott Parsons's work to the evolution of Bali's irrigation network. The differences all hark back to the kind of questions researchers want to answer. Frenken (2006a, 2006b), for instance, uses the generalized fitness landscape approach, with reference to Simon (1969) and Altenberg (1997a), to model complex technological systems in which interdependence in the operation of constituent elements forms the system. His generalized model is a model of technological design in terms of configurations, with which he runs simulations to see how transaction costs, decomposability, modularity and vertical disintegration should be integrated in future modelling exercises. Compare such an investigation with the work of for example Dooley et al. (2003) in their high-resolution broadband discourse analysis. They define fitness of texts by the words present in a text. They then examined 50 highly diverse texts to tentatively conclude that discourse evolves to the edge of chaos, which among other things could imply that speakers structure their discourse unintentionally in such a way that key terms are placed where they maximize their influence.

While both types of analysis are structured in fitness landscapes, the authors pursue entirely different research agendas in entirely different domains using different methods. It shows that fitness landscapes are heuristically open-ended, which invites scientists to play around with the concepts and to generate novel ideas and insights – something we will do in this book too. Criticism as voiced by for example Kaplan (2008) has its merits in so far as fitness landscapes are indeed half-proven, half-conceptual theories. However, as relayed above, this double-sided property doesn't need to be a problem as long as authors are aware of their own position vis-à-vis the framework. On top of that, we should remind ourselves that theories are rarely final in the social sciences.

We structured the diversity into five distinct modes of inquiry. In using the framework in non-biological domains, researchers have to negate certain trade-offs. Modelling fitness landscapes, the most popular adaptation, allows the researcher to stay close to the original *NK* model. However, working with real-world data, quantitative and perhaps even qualitative, makes it much harder to deploy exact copies of Kauffman's versions. Consequently, such applications are heavily simplified. For example, they use fewer variables, remain static and control poorly for exogenous factors.

Also consider the visual representation of two- and three-dimensional landscapes, which are the most accessible point of entry and offer scope for (case-based) narratives. However, a representation on three axes, of which the z-axis represents fitness as derivative of the two others, leaves exactly two variables to be represented. Many authors have found a way around this by using composite variables, but it still seems that the three-dimensional representation has severe restrictions in comparison to the generalized *NK* model.

The trade-off between generality and specificity matters for the purpose of this book, because one of the main issues in fitness landscape modelling is whether such models can be generalized to the extent that they can account for all variety, similar to what Altenberg (1997a) set out to do in biology in order to facilitate any number of elements and functions. We've seen that authors deal with this by developing models that are not calibrated or validated using real-world data or by engaging in case narratives that may be true for that specific instance but perhaps not across cases. For the purpose of this book it is important to reduce some of the ambiguity described above and not to avoid the persistent difficulties in modelling fitness landscapes. We should be particularly careful in addressing the many gaps between reality and model.

2.6 PREREQUISITES FOR MODELLING FITNESS LANDSCAPES

We have taken the broad view in this chapter, presenting a panorama of evolutionary theories in biology and the social sciences, and discussing the origin, development and diffusion of fitness landscapes in both domains. Throughout, we have argued against restricting such theories and models to one domain. As evidenced by the many authors discussed here, there are good reasons to keep working on cross-overs. To further the research programme of fitness landscapes in the social sciences, we need to take into account the variety, strengths and criticism discussed above. Above all, we need to bring something new to the table. We therefore follow the example of for instance Axelrod and Bennett (1993) and use the literature review presented in this chapter to derive five prerequisites that will serve as our guide when developing our own fitness landscape for collective decision making. Prerequisites can appear to be unnecessarily restrictive. This is not the case. Such intellectual coordinates are in fact very helpful in pinpointing the areas where scientific progress can be made and in navigating the pitfalls and weaknesses discussed in this chapter.

- The first prerequisite is that the model should be rooted in an unambiguous and consistent ontology and epistemology in order to avoid any confusion about the kind of causality that is assumed to drive reality and about the kind of statements that can be derived from the empirical findings. This prerequisite is in response to the observation that authors either are seemingly unaware of such matters or believe that science only comes in one flavour.
- Second, the model should be able to process data obtained from real-world as opposed to for example simulated data, because we want the model to be an accurate representation of the real world in order to fully account for the complexity of this reality. While we fully agree with the intellectual clarity presented in models and in the modelling process, we also want to move beyond the stage of prototyping fitness landscapes for the social sciences. This requires us to work with real-world data.
- Following the previous two points, the model should allow for unambiguous measurement in order to avoid confusion about what is being measured with the model and how it is being measured. We appreciate that measurement in the social sciences is ambiguous by definition, but that should not prevent scientists from at least operationalizing things.
- The fourth prerequisite is that the model should uncover persistent patterns between interaction and fitness in collective decision making. We came across quite a few studies in which such patterns were presented. Unfortunately, these models always used simulated data; that is, they violated the second prerequisite. It seems that the combination between the second and the fourth prerequisite is rather rare.
- The fifth and final prerequisite builds on the previous point and holds that the model shouldn't be too simplified or generic or generate obvious statements and truisms without additional depth. Naturally, any model of reality is a simplification – that is what models are supposed to do. But there is a point where simplification gets in the way of saying anything meaningful. If authors start their modelling attempt with page after page of simplifying assumptions – and some do that – then the model is no longer going to be informative.

These five prerequisites incorporate our findings about the possibilities and restrictions from the literature survey into our attempts in this book. This implies that modelling fitness landscapes is not just about mathematical elegance but that external validity is as important as internal validity. Unlocking the full potential of fitness landscapes in comprehending the complex dynamics of collective decision making requires that all these aspects are fully addressed. We appreciate that these prerequisites will be exposing the vulnerabilities of our own attempt, and we risk punching above our weight. We accept that. It is part of the scientific endeavour to try out new things. By definition, that implies that sometimes things should have gone in a different direction than they actually did (luckily, we have a very large rubbish bin in the office). We start this endeavour in Chapter 3, where we discuss our particular take on the epistemology, ontology and methodology of using fitness landscapes in understanding collective decision making.

3. The transformation of fitness landscapes

3.1 A COHERENT, UNAMBIGUOUS MODEL?

Now that you have worked your way through the origins, evolution and diversity of fitness landscapes, a simple 'It all depends' will not suffice as an answer to the diversity noted in the previous chapter. But neither does Kaplan's (2008) complaint that this diversity is a trace of people getting it wrong. This chapter therefore asks the simple question: what are fitness landscapes? Simple questions have a tendency to require complex answers, and this one is no exception. It forces us to take a much closer look at the ontology and epistemology of fitness landscape inquiries in the social sciences. As we have seen in the previous chapter, there is relatively little coherence in the many ways in which fitness landscape models are interpreted and applied, so we will need to get a better understanding of why this is the case. Having done that, we will then zoom in on our own position vis-à-vis the interpretation and application of fitness landscape models in understanding collective decision making. This position will direct us to a fitting methodology. As could be gathered from the previous chapter, we have already restricted ourselves by defining five prerequisites, but we will do our best to show the logic behind our choices.

3.2 CONTRADICTORY INQUIRIES

Let's return to Kaplan's issue for the moment. One could adopt the evolutionary position that the best version of fitness landscape models is the one that survives corroboration. Exactly which version survives in the long run can't be predicted in advance, so the only way to evaluate different models prior to the actual empirical test is to verify the logic of their methodological constructs rather than the outcomes or various other aspects (Watkins, 1989). Reasonable though this may sound, and fitting though this may be in a book about evolutionary theories, it avoids some persistent issues that need to be addressed. For starters, different theories in the social sciences can serve entirely different purposes and are therefore constructed in mutually diverging ways. Although one could still evaluate the methodological construct of each individual version, comparing them within the whole research programme wouldn't make any sense owing to the different purposes for which they have been developed. Try comparing, say, Levinthal's model for simulations (1997), with which he modelled firms with N attributes and K epistatic interaction effects to explore the implications of adaption and selection, to Rhodes's attempt to map the case studies in urban regeneration processes to fitness landscapes in order to explain the emergence of new structures (Rhodes and Donnelly-Cox, 2014). There is no sensible way to tell which one is better, because the first presents models using simulated data in order to explore certain algorithmic effects whereas the latter develops case studies to explain certain empirical outcomes in the field of urban planning. They serve such different purposes in such different ways that they are comparable only *within* their respective modes of inquiry, that is, in comparison to other similar approaches to fitness landscapes. However, these subsets are too small to be made comparable in any sensible way. So this approach doesn't really solve the question of which one version within the whole research programme about fitness landscapes is better.

Bartels (1987) in his overview of the fundaments of the philosophy of science asserts that it is logically impossible to prove unambiguously which particular version of a theory is a better one, because selection of a given theory is not necessarily a scientifically rigorous process. Theories may be rejected or accepted by chance, and there is ample evidence in the history of the social sciences that certain theories were first disregarded and then rediscovered later, such as in the case of systems theories (Gerrits, 2012). As we have argued in the previous chapter, fitness landscapes are not different in this regard (see also Plutynski, 2008). Moreover, falsification only works if the target theory is geared towards testable propositions. If not, this criterion falls short. Social scientists produce all kinds of theories, not just mathematically precise models and testable hypotheses. There are theories that offer grand ideas, theories that suggest alternative ways of understanding social reality, and many more variants. As we've seen, many authors use fitness landscapes as a metaphor or as a means of sensemaking, which taps into a different, non-positivist aspect of social scientific inquiry. That is, metaphors and sense-making serve as a means to construct new ideas or raise new questions, rather than featuring causal explanatory power (Chettiparamb, 2006; Lakoff and Johnson, 2003). Normally, explanatory power serves to make both the subject and the scientific practice intelligible (Woody, 2004). But this doesn't mean that metaphorical uses are less valid as a mode of inquiry than, for instance, modelling and simulation attempts. In these attempts, the original fitness landscape model is reconstructed to be applied to new targets (Peschard, 2011). The model and simulation efforts constitute epistemic tools for gaining information about the model itself, enabling users to draw conclusions about the target system (Bolinska, 2013). All ways of inquiry discussed before have a particular contribution to make to the body of literature; that is, they all have added value depending on the questions to be answered. 'Representations meet the epistemic aims of some domains and can come to shape these aims. As a science makes progress, the representational framework determines which questions are worth asking and how one should go about investigating such questions' (Plutynski, 2008: 621). To this we add our point already made in the previous chapter: theories in the social realm are rarely 'final', as researchers keep refining, amending and extending them. So, while Lakatos (1976) may be right in suggesting that natural selection will decide which version will come out on top in the long run, we need to accept that this doesn't help us in the present if we want to decide which versions will be most helpful given a specific research question.

If long-standing practices of social sciences consistently fail to meet the criteria of scientific purity we may have to come to terms with the idea that those principles are misguided. Given the heavily contingent nature of social reality, it may very well be that there is simply no instance available with which one can assume falsification of a given theory. The best we may achieve is detecting that something is different from a previous situation, which is not even close to falsification (Sayer, 2000). Falsification relies on closed systems, and social reality is anything but closed (ibid.). We therefore need to question Popper's (2002) assumption of scientific purity that lies behind the idea of corroboration. Perhaps plurality in fitness landscape inquiries is not a shortcoming but a feature inherent to the social sciences. The different modes and the types of statements generated within these modes should not be judged as separate competing research programmes but as inquiries constituting one, albeit diverse, research programme. Can such inquiries have a dual, perhaps even contradictory, purpose inside a research programme?

At this point, it is useful to borrow Althusser's classification of social theories into Generalities I, II and III (Mouzelis, 1995) in order to explain the coexistence of contradictory components within one programme. Generalities I concern the raw materials and grand ideas that, while untested, constitute coherent wholes that break the boundaries of the status quo in science and can be considered to have a paradigm-shifting impact. They arguably are research programmes in themselves. Generalities II concern theories as conceptual approaches that suggest novel ways of looking at reality within a given research programme, in other words theory as a tool or means to further social inquiry. Such approaches are historically contextualized; that is, they are informed by what are considered current persistent issues, and may very well contain contradictions. Note the difference with Generalities I, which use a similar device to contest the implicit assumptions that are dictated by contextualization. Generalities II operate within that accepted context. Generalities III concern theories as provisional end-products that generate substantive claims that can be falsified. As such, they may be decontextualized, because they point at certain causal patterns that are considered general or universal until a better version has been proposed and tested. Assessing theories first and foremost means understanding what kind of theory one is dealing with. According to Mouzelis (1995), Generalities II theories in the social sciences have met considerable criticism for being abstract and non-testable. Regarding the ways in which some authors write about fitness landscapes, we can relate to the observations about abstract and non-testable statements that rely heavily and solely on the readers' sympathy for their acceptance. However, we also fully agree with Mouzelis that such criticism is unfair, because those theories were not meant to be testable in the first place. In his view, Generalities II theories are supposed to 'construct sets of conceptual tools . . . for looking at social phenomena in such a way that interesting questions are generated and methodologically proper linkages established between different levels of analysis' (1995: 3–4). Arguably, it is the *dialogue* between Generalities II and Generalities III that is necessary for the advancement of social sciences rather than one *or* the other. The conceptual tools in Generalities II offer avenues of thought, while the substantive tools in Generalities III probe the possibilities of explanation. Coming full circle, the findings from the probes inform conceptual approaches.

We are the first to admit that we use Althusser's classification in a pragmatic fashion. Althusser originally developed it in order to make sense of the nature of Marx's writings (not to be confused with the works of one of the authors of this book!), which resulted in texts not much easier to understand (Kurzweil, 1980). However, the heterogeneity in fitness landscape research suddenly starts to make much more sense from this perspective. It should be stressed that the duality between Generalities II and III – between conceptual narratives and testable propositions – is also present in the work of Kauffman, who never claimed that each statement was already tested or even testable (Castellani and Hafferty, 2009; Rosenhead, 1998). The same goes for Wright's work that, despite its apparent stern formality, matched pieces of proven theories with more tentative suggestions for possible explanations. In testing, not every element from Wright's work turned out to be correct. For example, under certain conditions selection doesn't drive populations on to higher peaks but into valleys instead

(Plutynski, 2008). Petkov even goes as far as saying that the term 'adaptive landscapes' or 'fitness landscapes' is first and foremost an umbrella term that encompasses the metaphor, a metaphoric vocabulary used to interpret the mathematical models, and the graphical representation that illustrates some of the features of the model. They serve 'as a general unifying conceptual framework, which has greatly facilitated the synthesis and which still provides a basis for unifying heterogeneous evolutionary phenomena and explanations' (2014: 3). Consider that, 'since there exist at least two variants of the metaphor that can be expressed with the same graphical representation and discussed in the same terms . . . and those variants are not completely mathematically equivalent, the metaphor taken by itself leads to a confusion' (ibid.). Kaplan (2008) seems to agree, noting that the metaphor alone could play different roles, depending whether the diagram or the three-dimensional visualization is taken as the point of departure. Something similar applies to applications in the social sciences. Each of the authors we evaluated would select and highlight certain elements and not consider other elements, all in the light of a specific research question.

Let's return to our initial observations and questions: do ambiguity and contradictions mean that fitness landscapes are in a state of theoretical anarchism where anything goes until some versions give way to evolutionary pressures? Plutynski (2008) traces the differences in the literature in biology back to the dichotomy discussed above: testable hypotheses and propositions, and the proverbial conceptual dictionary for interpretation and sense-making coexist within the totality of fitness landscape research. These two may be contradictory at a given stage of inquiry, but they are not mutually exclusive, as Althusser helps us understand. A set of testable propositions that may point at a (general) occurrence is not yet an explanation. One needs the dictionary to make sense of the pattern that has been detected in order to generate an explanation. That, in turn, informs new testable propositions. In other words, contradictions are not necessarily problematic but a marker of the extent to which a theory has developed. As such, they have a dual purpose that is essential for the development of the research programme as a whole. For example, using fitness landscapes as metaphor or as a tool for sense-making means telling narratives about how actors arrive at decisions and how it can be understood that certain actors may accidentally find or lose a fit with their environment. While such stories are very far removed from testable hypotheses, they still suggest new possible ways of understanding collective decision making, for example in terms of an adaptive walk in search of a better fit. We noticed how such stories, despite their seemingly scientific sloppiness, were able to render vivid images. And, even though theorizing can be perceived as less accessible, it still has the power to open new ways of thinking about collective decision making (e.g. Beinhocker, 2006). Case mapping can be considered a mode of inquiry that fits better with Generalities III. By using fitness landscapes to structure and relate empirical data, the researcher attempts to arrive at causal patterns. These patterns are open to some form of falsification (though not in any sense Popper-like), which in turn means that they can inform the conceptual side of fitness landscapes. Obviously, modelling is also a clear example of Generalities III, because it renders testable propositions and provides the means to test them, either as simulation or against real data. One way or the other, the various kinds of inquiry inform each other, sometimes explicitly and sometimes implicitly, sometimes in a structured way, sometimes in a way that is a little unstructured.

3.3   MOVING BETWEEN BIOLOGY AND THE SOCIAL SCIENCES

We have now established that the coexistence of different versions of the same theoretical framework is perfectly acceptable, and even necessary if a research programme is to advance. Underneath this discussion of the state of fitness landscape research in the social sciences lingers a second issue, namely whether theories and models from biology address the same causal constructs and generate the same kind of explanation in the social realm. While we agree that evolutionary theories at large have a role in the social sciences (cf. Byrne and Callaghan, 2013), we will argue here that one also needs to be constantly aware of the need to transfer and operationalize those theories and models for the specific characteristics of the social world. In our literature survey, we found that many authors simply took Kauffman's propositions to be as true in the social realm as they are (or could be) in biology. This conveniently ignores the fact that the properties of the social realm are different from those of biology. An obvious difference, yet rarely addressed by authors, is the fact that actors in the social realm engage in *conscious* and *intentional* decision making, for which they (attempt to) forecast, anticipate, plan and reflect. They show emotions, do unexpected things or do nothing at all. In other words, there is consciousness at work when making decisions. This is very different from genotypes or phenotypes searching for better fits. These don't have reflexive capabilities, and their adaptive moves are always based on *past* performance instead of on anticipation of the future (cf. Kauffman and Levin, 1987). Many other such differences exist, some clear, some more subtle. This is not to say that theory transfer doesn't make sense – see for example Luhmann's (1984, 1995) influential use of biological autopoiesis

(Maturana and Varela, 1980; Varela, 1981) in social systems – but rather that transfer is an elaborate process that can't be sidestepped.

To make a distinction between biology and the social sciences is to upset a certain crop of scientists who believe that, ultimately, *all* explanations will, or should, follow the same logical structure. This harks back to Mill's understanding of causality and generality and has permeated the natural as well as the social sciences. We argue that this claim is wrong as far as the social sciences are concerned. Yes, many and perhaps all social scientists are aware of the fact that a certain causal construct in the natural sciences doesn't necessarily account for social complexity. Yet, as among others Abbott (2001) has argued, many accounts of social phenomena have a tendency, at least superficially, to discuss such social complexity as a dehumanized construct with variable-based language and a focus on the relations of (a limited and stable set of) variables instead of focusing on people doing things in certain contexts. This is simply a matter of convenience: 'Action is reality. But familiarity and practice send the caveats packing' (Abbott, 2001: 98) and leave us with much research that hinges on dehumanized variables. Winch (2008) traces this tendency back to the rise of the social sciences in the 1950s and the subsequent desire for the social sciences to mature as a natural science, complete with a rather simplified understanding of social causation, attempts to define general laws governing the social, and aspirations to generate subsequent predictive power. But the simple matter is that social complexity is *different* from natural complexity. While

> human reactions are very much more complex than those of other beings, they are not *just* very much more complex. For what is, from one point of view, a change in the degree of complexity is, from another point of view, *a difference in kind*: the concepts which we apply to the more complex behaviour are *logically different* from those we apply to the less complex. (Winch, 2008: 68, italics added)

Let's use an example to illustrate this point. Kauffman speculates that 'species can tune coevolution not only by tuning the ruggedness of their own landscapes, but also by tuning how many other species impinge on them' (1993: 263). In other words, when species are connected, they can manipulate the reciprocal relationship with other species. Intuitively, this could also be true for social actors. After all, it doesn't take much imagination to understand that humans can decide where, when and how to connect. But then the complexity starts to multiply quickly. Interactions in the social realm can take on very different qualities, for example the difference between meeting at the coffee machine or in the courtroom. And by what means do humans actually connect or disconnect? Emotions and perceptions play an obvious role in this: love, dislike, passion, regret, pride, sadness, enjoyment and many others guide human actions. There are no formal solutions for such dimensions within the fitness landscape framework. Or consider the fact that the same people may be simultaneously connected and disconnected, for example as business partners but not as friends. When we, not unreasonably, add context and time to the mix, the matter quickly becomes very complex indeed. It is equally reasonable to assume that the research into genotypes doesn't need to worry about whether a gene was sad or happy today, or whether its relationship with neighbouring genotypes is 'complicated'. For social scientists, observing behaviour itself is not enough. One is required to understand why people do certain things. 'Learning what a motive is belongs to learning the standards governing life in the society in which one lives; and that again belongs to the process of learning to live as a social being' (Winch, 2008: 77). The apple will always fall from the tree, but humans won't, because they have learned that gravity hurts. One needs to take that reflexive layer into account.

The matter also boils down to ways in which human behaviour is explained, that is, the kind of narrative that is given to the patterns detected in research. If anything, the postmodernist turn in the social sciences has taught us that scientific explanations are heavily context-dependent and not self-evident from an objective theoretical construct. It is the researcher who assigns meaning to the patterns detected. Our overview of evolutionary theories in the previous chapter has already demonstrated that this is the case for both biology and the social sciences. In order to say something meaningful about the social and in order to understand how the explanation came about, one has to have knowledge of the particular context in which the explanation is generated. As Winch has already pointed out, 'the concepts and criteria according to which the sociologist judges that, in two situations, the same thing has happened, or the same action performed, must be understood *in relation to the rules governing sociological investigation*' (2008: 81, italics in original). Indeed, one could argue that there are many ways of making the world more intelligible: through the theories and constructs of the natural sciences or through the social sciences, but equally through the humanities and arts. Each mode has its own rules for doing so, tailored to what is deemed important. It is also from this understanding that it doesn't make much sense to assume that a model from biology renders, by definition, equally valid explanations in the social sciences.

Let us stress again that this is not a diatribe about the differences between biology and the social sciences. Indeed, we assume an anti-dualistic position regarding the physical, the biological and the social (Newton, 2003, 2008). In essence, we simply argue that one needs to account for the different properties in the source and target domain when engaging in theory transfer. Indeed, all types of inquiries have their merits and pitfalls, as corresponds with the debate on how any representation can accurately describe the world (Knuuttila, 2011). Of course, the question is what kind of representation is achieved. Structural representation allows drawing conclusions about the thing it represents, while surrogative reasoning allows using one sort of thing as a surrogate in our reasoning about another (Swoyer, 1991: 487). Models such as fitness landscapes can be regarded as vehicles for surrogative reasoning when used in the social sciences, that is, 'reasoning from premises in a vehicle to conclusions about its target system' (Bolinska, 2013: 219). The danger in a strict application is that idealized models cannot accommodate certain features of their target domain, while a rather associative or looser application could mean that any two things can be regarded as arbitrarily similar (ibid.: 220). Another side-effect is that (representational) models can create new models in the target domain; that is, they have a generative constructive use (Peschard, 2011). The core of the matter is that any representational model should achieve 'articulated awareness of the nature of the objects and relations constituting that particular domain' (Woody, 2004: 782). Of course, this requires that the researcher has

> reasonable knowledge of both the source and the target domain, sufficient to enable a pertinent abstraction of key relational characteristics from within each; an effort to draw out and explicate key similarities and analogies; an effort to abstract and elucidate essential relational features, and also an attempt to explore the abstractions with relation to other theoretical work in the target domain. (Chettiparamb, 2006: 78)

Without these, the transfer can't constitute anything more than a literary metaphor that can be used for 'persuasive purposes, or as a gap-filling device for inadequacies of language' (ibid.), but nothing more than that. The issue is further complicated when we take into account that there is a difference between what is similar in both domains and what *appears* to be similar but in fact isn't. There are always properties in the source domain that are denoted in the transfer but are not present in the target domain. Yet the transfer carries over the connotations from the source domain, which can enhance the transfer but also lead to a wrong understanding of how the theory operates in the target domain.

In short, we are dealing with a specific kind of theory transfer in this book – from evolutionary biology to issues of collective decision making – that comes into five distinct modes of inquiry and that plays specific roles in generating knowledge, according to Althusser's generalities. Naturally, we will have to make choices – ontological, epistemological and methodological ones – in order to explain the kinds of statements we generate in the empirical part of the research. We think critical realism offers an avenue for this.

3.4   A CRITICAL REALIST APPROACH TO FITNESS LANDSCAPES

Let's get one thing out of the way first: the standard positivist approach to the social sciences is dead. No, seriously. Let's revisit Abbott's critique of positivism (2001), namely that it assumes that: (1) social reality is made up of fixed entities with varying attributes, commonly expressed in the language of variables; (2) cases operate independently, that is, what happens to one case doesn't constrain or enable what happens to others temporally or spatially; (3) attributes of cases have only one causal meaning; and (4) such attributes determine each other as independent scales rather than as constellations or configurations of attributes. Cases in the social sciences don't work that way. They are heavily contextual and evolve over time, sometimes even to such extremes that 'a case which began as an instance of one category may complete a study as an instance of another; a state can become a nation, a craft can become a profession, and so on' (Abbott, 2001: 142). This means that we can't rely on the mechanistic worldview we touched upon in the first chapter. However, it doesn't mean that everything has become fluid and random. Indeed, cases are often connected and do exert mutual influence. They are produced as configurations of factors rather than as a set of discrete variables, let alone variables that are either independent or dependent. In addition to Abbott's points, we should point out that the main thrust in positivism is to develop general laws, that is, causal laws of natural necessity that apply to all similar cases and that can be upheld regardless of time and place. As it turned out, laws in social reality are either trivial (see for example Abell, 2004's comments) or notoriously hard – or even impossible – to grasp. The fact that we fail to detect such laws in human behaviour is not a weakness of the social sciences but rather a sign of the complexity of social life (see for example Green, 2015 on a similar discussion within biology). The final nail in standard positivism's coffin is that the fact–value dichotomy that underlies such positivism has been successfully debunked (e.g. Fischer, F., 1998, 2003 in the realm of collective decision making). As we have argued above, meaning and behaviour form a coupled social construct that defies the analytical power of positivist reductionist research instruments.

In short, let's not talk about standard positivism any more. Equally, however, this is not meant to be a call for complete relativism and the kind of word-play that came with the postmodernist turn. Postmodernism had utility in pointing out the weaknesses of positivism, but somewhere along the line it became a semantic game that does not really inform the types of questions we are asking. Nonetheless, to say that these differences led to distinctively separate kinds of science is grossly overstating things. Some of the core assumptions of positivism in the natural sciences, such as time symmetry, have for some time also been under critical scrutiny, in particular from the perspective of the complexity sciences (Prigogine, 1997). Indeed, time asymmetry impedes general laws in evolutionary biology (Mitchell, 2009). Conversely, the social sciences fall short when always addressing any observed phenomenon as a social construct.

The ontological point of departure is therefore neither positivism nor post-positivism but a form of complex realism that mediates between the two (Bhaskar, 1979, 2008; Byrne, 2002; Reed and Harvey, 1992; Shotter, 1992). Critical realism negates that divide between the natural sciences and the social sciences, between relentless positivism and overly relativistic post-positivism (Wuisman, 2005). It accepts that there is a reality outside our perceptions that can be known through the deployment of scientific tools (Losch, 2009), as such echoing Kant's empirical realist stance. Critical realism holds that using such tools can generate pertinent insights (Sayer, 2000). However, causal relations are deemed to be contextual; that is, they are local in place and temporal in time. They can be known, but the resulting knowledge is contextual by definition (Byrne, 2005; Cilliers, 1998). It is not necessarily positivism's antidote but rather a much more refined version (Byrne, 2002) that allows the convergence of fact and value in the analysis of complex causation.

Within the philosophy of sciences several somewhat different strands carry the label of critical realism (Easton, 2010; Losch, 2009). We specifically build on the work by Roy Bhaskar (e.g. 2008), who acknowledges the existence of a knowable reality and posits that the mechanisms underlying social events can be uncovered in order to gain an improved but still contextualized understanding of that reality (Cilliers, 1998, 2002, 2005; Gerrits, 2012; Gerrits and Verweij, 2013). Bhaskar deals with the fact–value dichotomy mentioned above through a stratified understanding of reality. He distinguishes between the world and the perceptions with which this world is viewed but also between the real, the actual and the empirical (Sayer, 2000). In this view, mechanisms that can cause events to happen are potentially always present but only activated in certain situations; that is, the mechanisms are conditional. Events and processes in the real world (the second dimension of stratification) and the effects of the mechanisms (third dimension) can be observed and researched as they take place (Gerrits and Verweij, 2013). Compare this to Kauffman's conceptualization of self-organization as a causal force. Kauffman can explain the mechanisms and logics of self-organizing processes by introducing the theory of adjacent possibilities, as already discussed in the previous chapter. Adjacent possibilities fit perfectly with Bhaskar's concept of stratified reality, because they rest on the same understanding of unfolding processes: a distinction between what *is* (i.e. actualized states), what *will be* and what *can be* (i.e. potential that can become unleashed under certain conditions or by certain mechanisms). The important factor here is context, that is, the specific environmental circumstances that influence certain processes and that can trigger bifurcation in speciation (cf. Gavrilets, 2003). Critical realism pivots on that aspect. Logically, there is no room for decontextualized cases in this view.

From the contextualized take on reality it follows that this reality is equifinal and multifinal; that is, different conditions can bring about similar outcomes (equifinality) or similar conditions bring about different outcomes (multifinality). Again, a critical realist view has no issues with localized causality but refuses to suggest that general, decontextualized laws can be identified. This also applies to biology. After all, each evolutionary progression is the irreversible result of a species adapting to a specific context (Mitchell, 2009). By and large, fitness landscapes allow the researcher to uncover equifinality and multifinality, to show which mechanisms have been at work during collective decision making and to generate an explanation for the dynamics of the landscape of a particular case. Fitness landscapes allow for a contextualized understanding of how events take place and how actors interact. Therefore, they sit comfortably with a critical realist understanding of social reality. For the present purpose, we want to focus on four focal points based on critical realism that are key for our research into collective decision making and the use of fitness landscapes in the analysis of such decision-making processes: emergence; time; context and configurations; and action and interaction (see also Gerrits and Verweij, 2013 for an in-depth discussion).

3.4.1 Focal Point 1: Synchronic Emergence

Central to our critical realist understanding of social reality is the concept of *emergence*. Note that we use the term here not as a general catch-all term or to denote a somehow magical appearance of the world, but rather as a specific complexity-informed way of understanding how social reality unfolds. Probably the most well-known version of emergence can be found in Holland's (1995, 2006) work, which, in popular terms, says that the systemic whole is more than the sum of the parts constituting the system. Various computational simulations have demonstrated how complex structures such as flocks can emerge from relatively simple rules of behaviour in agents (Langton, 1986; Reynolds, 1987). Collective decision making can be, and often is (Axelrod, 1984, 1986, 1997; Schelling, 2006), seen as a truly emergent phenomenon when actors, acting upon their own particular set of goals and behavioural rules, create intricate patterns and outcomes in terms of for example institutions. Emergence is also very much present in Kauffman's work because of its focus on self-organization. Indeed, self-organization concerns the emergence of a certain structure without planned and superimposed design. Adjacent possibilities revolve around the idea that novel processes and structures can emerge past bifurcation points. Thus emergence is inherently tied to evolution.

Emergence draws attention to the fact that social reality is a systemic whole that is not fully explicable in terms of that from which it emerges (Byrne, 2001), and whose integrity will be violated when it is taken apart in discrete elements (Gerrits and Verweij, 2013). However, the most common method of researching emergence rests on the premise that there is such a thing as a starting point and final outcome, that is, that the starting conditions are set and that emergence can be traced from those starting conditions. This is usually demonstrated through simulations (e.g. in agent-based models) where one can indeed define the precise starting conditions prior to the simulation. A diachronic take on emergence is at odds with our second and third prerequisite. Such conditions are lacking in social reality (Elder-Vass, 2005). Diachronic emergence, in which parts and wholes are set apart in time, is therefore heuristically interesting but will pose problems with measurement with real-world data and with using any means other than simulations. Social reality is characterized by synchronic emergence, where the temporal division between parts and whole is brought back into one single but continuous instant. Thus emergence is not a thing out there that can be measured as done in simulations – it is a way of understanding how reality unfolds.

As for example Byrne and Elder-Vass have explained, the consequence of synchronic emergence is that reality is non-decomposable. Non-decomposable reality can't be described in discrete components in any meaningful way, because structures and processes come about through internal and external interaction. Taking them apart loses the aspect of interaction and thus its explanatory value. We therefore need to treat cases as systemic and emerging wholes that can develop in similar or dissimilar ways under certain conditions, that is, tracing their equifinal or multifinal nature. Accepting emergence as the primary driving force of social reality, as we do, has implications for the use of variables, which we will demonstrate further below.

3.4.2 Focal Point 2: Time and Event Sequences

Emergence also focuses our attention on time (e.g. Byrne and Callaghan, 2013) because it is a process. Time-bound dynamics are important in both the source domain and the target domain (Newton, 2003), as evolution simply *is* time and so is social reality. It is this dimension that takes fitness landscapes out of the realm of fixed configurations and comparative statics. Time helps in arranging or ordering social scientific research, because 'social reality happens in sequences of actions located within constraining or enabling structures' (Abbott, 2001: 183).

Time has three properties that needed to be taken into account in the research for this book. The first is that it is asymmetric or unidirectional. At any point in time, there are *n* possible future directions. Given the specific configurations at that point in time, some of these possible futures are more likely than others. On top of that, there is always an element of chance – the novel conjunction of previously separate events – that can open up unforeseen new future directions. The world is, to borrow the terms of Prigogine (1997), developmentally open because of the possibility, however small, of chance. Obviously, this means that purely linear conceptualizations of time don't hold much explanatory value. Second, time asymmetry impedes the ability to foresee and predict. A meaningful analysis encompasses a retrospective reconstruction of how social reality unfolds or emerges from antecedent situations – as mentioned before, these statements all hark back to emergence as the most fundamental aspect of social reality. We can tell what has happened and why it happened but not whether it will always happen again (Byrne, 2002). However, do note that this particular treatment of time doesn't mean that the world is continuous and constantly in flux: it is punctuated. This is the third property. As shown through for example punctuated equilibrium theory in biology (Eldredge and Gould, 1972; Mayr, 1963), changes or progressions are asymmetrically distributed over time units. The same dynamics, where periods of stasis are punctuated with bursts of swift change, have also been observed in the social sciences in general (Elster, 1976, 2007), and in collective decision making in particular (Baumgartner and Jones, 1993). 'Much of sociality . . . can be imagined as a structure in which actors proceed through trajectories [those parts of the process that aren't bifurcation points] to their ends, then face the striking and to some extent randomizing moments of turning points' (Abbott, 2001: 25). Certain standard methods may very well work with the analysis of stasis, but one needs sequential methods to analyse the occurrence and effects of bifurcation, to see how potential states become actualized. Social complexity can only be researched in a meaningful way and understood if the time dimension is included in the analysis (Byrne and Callaghan, 2013; Gerrits, 2008, 2011, 2012).

The three properties of time mentioned here, asymmetry, impeded foresight, and punctuation, mean that our research should not only be longitudinal but also feature a high resolution and identifiable time units in order to explain the shifts in the landscape. Abbott's proposal for a time-centred approach to the social sciences, in which actors and structures are primarily defined as *networks of events*, operationalizes how time can be put to analytical use in cases. He is particularly interested in the unfolding processes in which certain events are enchained, and how sequences of events provide explanatory power for the coming-about of certain phenomena. The ways in which events occur, in particular in precedent–antecedent patterns, become the analytical focus. Sequential methods could uncover regularities in such sequences of events, including the occurrence of bifurcation points that require at least two points in time for their identification. Abbott arranges related events in so-called lineages. Such lineages may diverge, converge or intersect. Events can be members of multiple lineages simultaneously and can therefore exert influence on various lineages even though they may not be of pivotal importance in each lineage. For the researcher, this means that 'Every event has multiple narrative antecedents as well as multiple narrative consequences' (Abbott, 2001: 192; see also Poole et al., 2000; Spekkink and Boons, 2015, for a similar approach). Compare this to Kauffman's coupled landscapes in which the adaptive moves of genotypes (i.e. changes in response to selection pressure) in one landscape can influence the properties of neighbouring or coupled landscapes. Abbott proposes the same kind of coupledness. Naturally, that poses a number of analytical problems, in particular that 'most things that could happen don't' (Abbott, 2001: 18) or, in other words, that we can imply the occurrence of bifurcation points but we can't logically say anything about the properties of the non-actualized state. More simply put, something else could have happened at a given point in time, but it didn't, so we will never know what could have happened. The second analytical problem is that the selected time horizon will influence the type of statements researchers can make. The time span observed can give the impression that a process is excessively dynamic or completely static, while a change of the time frame could lead to different conclusions if that extension or exclusion features a static or a dynamic section.

There is also an issue with how changes in time can be represented. We can certainly log changes, or lack thereof, longitudinally in graphs if we work with quantified data. But, as we explained above, that doesn't *explain* why the changes occurred. In addition to quantitative measures, we need qualitative descriptions with which transitions can be made sense of. Rich narratives are essential means for mapping and understanding transitions or shifts and the conditions under which they occur (see Abell, 1984, 2004; Vayda, 1983, on progressive contextualization). Abbott refers to such descriptions as narrative positivism, to indicate that such qualitative, thick descriptions complement but do not substitute for quantitative measures. The most important feature of narrative positivism 'is that narrative meaning (the "causal force" of enchainment) is a function of present and past context' (Abbott, 2001: 193). It is up to the researcher to argue that certain events belong to certain lineages, the main criterion being that such events must be arranged in a plot that sets them in a kind of causal order on the basis of plausibility (or, in Abbott's own words, 'followability'; 2001: 290). Thus one has to pay considerable attention to explaining the beginning, middle and changes to the plot, all leading up to its conclusion. As with any method, a description is a reduction of the actual complexity, but such descriptions still do much better than just relying on a limited number of variables where social complexity is reduced to one dimension in the data space.

3.4.3 Focal Point 3: Context and Configurations

The multiple memberships of events in different lineages take us back to the point that social reality is heavily contextualized. As we have argued above, it is under certain conditions that state changes may be triggered. What we are dealing with here is the completely intertwined nature of process and structure through emergence. Thus social sciences are about 'those phenomena that are fully enmeshed both in social time and social space, what I have elsewhere called interactional fields. It is because we study interactional fields that we are a discipline of social *relations*, concerned with the social *process*' (Abbott, 2001: 124, italics in original). Structural properties in social reality are network-like (Byrne, 2005), which means that they are defined through interaction, in accordance with the thesis of emergence. As the environment changes, so does the system under scrutiny reciprocally.

Emergence and context mean that we need to understand each observed instance as combinations or *configurations* of events (see for example Byrne and Ragin, 2009; Ragin, 2014; Ragin and Amoroso, 2010, for an extended argument about configurations). We can trace the idea of configurations back to Wright's original work on adaptive landscapes. In 1978, he wrote:

> From my studies of gene *combination* . . . I recognized that an organism must never be looked upon as a mere mosaic of 'unit characters', each determined by a single gene, but rather as *a vast network of interaction systems*. The indirectness

of the relations of genes to characters insures that gene substitutions often have *very different effects in different combinations* and also *multiple . . . effects in any given combination.* (Wright, 1978: 1198, italics added)

These statements highlight that Wright was essentially thinking in terms of configurations and the net effect of such configurations instead of in terms of discrete variables. In addition, small population drift is a real effect that would otherwise be drowned out in the sheer numbers of large populations. That forced Wright to consider the possibilities of equifinality and multifinality in such combinations or configurations. Ragin follows a similar argument for the social sciences, which boils down to the simple fact that various aspects of cases can be evaluated separately from each other, in particular 'with respect to their "independent" causal effects on some outcome' (Ragin, 2000: 64).

Following Wright, Ragin and others, thinking in terms of configurations has far-reaching implications for the issue of whether the dimensions in the fitness landscape are independent or not. Independence implies that each coordinate on the grid is a possible empirical manifestation. But there are quite a few problems with that. First, given the nature of emergence, it is extremely unlikely that all configurations (i.e. all coordinates on the grid in the landscape) can occur empirically – some configurations are much more likely than others, and some are only theoretically possible. Second, and in terms of modelling, real independence of the landscape's dimensions would mean that a large distance between multiple actors could coincide with zero distance on any another dimension, which is impossible in real cases. Since we don't want to violate the second and third prerequisites, we need to accept that independence of variables in the landscape can only occur theoretically but not in reality. Consequently, we must substitute common thinking in terms of independent variables for thinking in terms of configurations. In fact, Abbott makes the same argument when it comes to the social sciences. We will elaborate on that point further below.

If we accept social reality as an emerging, contextual and non-decomposable whole, it follows that this reality should be studied as embedded or nested networks – very much in line with what for example Byrne and Callaghan (2013) and Byrne and Ragin (2009) recommend. We are aware of the fact that most of the adaptations of fitness landscapes in the social sciences follow a decontextualized, population-based approach, that is, where *n*-agents are relatively homogeneous and where the pay-off is unambiguous. We don't want to dismiss the usefulness of such an exercise, but we want to move beyond this and fulfil prerequisites 2 and 3. The primary reason for those two prerequisites is our uneasiness with considering human behaviour as a decontextualized and deindividualized phenomenon, as if humans are not subjected to the context in which they operate and as if information about a (relatively) homogeneous population tells something pertinent about the many complex considerations that real individuals have and the trade-offs they make. Given these points, a configurational approach is the only viable option.

3.4.4 Focal Point 4: Action and Interaction

If we accept that social reality is constituted in configurations, we still need to ask: what is at work here? Wright and Kauffman, as well as social scientists such as Abell, Abbott and Byrne, point to the importance of interaction, that is, the *conjunction* of the constituent elements that drives emergence. As Byrne puts it:

> Complex systems are to be understood not in terms of their parts, the analytical error, nor in terms of their wholes, the reserve holistic error, but in terms of parts, interactions among parts, wholes and the interaction of the whole with the parts. The word 'interaction' is vitally important. In reality, things work together and what they produce is not predictable from the inherent character of things themselves. Emergent properties contradict reductionism. (Byrne, 2002: 16)

Compare this to Kauffman's focus on interaction in fitness landscapes. Arguably, he points out exactly the same properties even though he may not deploy similar vocabulary. But his focus on interaction, as much as Byrne's approach, means an implicit shift away from the common approach of independent variables having an effect on dependent variables – even though Kauffman uses the language of variables to make his points.

Then what drives interaction? Human action, of course. It is human action and the motives behind it that we should focus on in an event-based, configurational social science. The argument is much older than for example Winch's (2008) argument mentioned earlier in the chapter. All theoretical reasoning aside, the motive is concisely summarized in the Thomas theorem, which holds that, when men define situations as real, they become real in their consequences; that is, humans interpret the world around them, which is a highly subjective assessment, and subsequently act accordingly. In the process, they create a reality that is based on their subjective understanding of the world they find themselves in (e.g. March, 1994 in the context of collective decision making).

3.5   BYRNE'S 'DOWN AND DIRTY EMPIRICISM'

The four focal points and our considerations force some methodological consequences. Any longitudinal structuring of events needs to be supplemented with a narrative that arranges the sequences in a plausible way. We also suggested moving away from variables and towards configurations of conditions. There is a clear relationship between narrative positivism and configurations (Savage, 2009; Savage and Burrows, 2007; Uprichard, 2012, 2013; Uprichard and Byrne, 2006), so let's dwell on these points for a moment. Both Abbott and Byrne point to the focus on variables as one of the biggest problems in researching social reality, because it takes the researcher away from actions, motivations and interaction. As stated before, everyday parlance in the social sciences has (unwittingly) substituted social reality in which people do things for a technical language in which variables do things. Variables are seen as representations, but if that is the case they seem to be doing a poor job. They shrink and flatten rich data to unidimensional concepts that can take on a limited set of (quantitative) values. In the words of Abbott, 'one of my longstanding objections to the standard approach is that human action basically disappears in it, to be replaced by the medieval shadowboxing of variables' (2001: 298). It leads Byrne to declare 'death to the variable' (2002: 31), which may be somewhat strong-worded but still to the point.

The debate also revolves around the issue of causality. The language of variables, the mathematical ways with which we can express, relate and model them, conjures up the image of firm, mechanical and general causality that is often – if not always – far removed from reality. Byrne calls for an empirical approach to understanding how the social comes about:

> The adulation of the formalized mathematical model, the assertion that the abstract representation of the world through establishing a causal model based on variables and isomorphic with an algebra – the construction of interpreted axiomatic systems – is precisely . . . abstraction over the real. We absolutely need a down and dirty empiricism in which understanding is grounded in the real and constantly returns to the real. (Byrne, 2002: 42)

Unfortunately, this point is often overlooked, because a formal model is simply more elegant to work with and, as some of our colleagues tell us, it *could* be true, right? Out of such considerations, a faux conceptualization of causation has crept in where it had become 'a property of mathematical and statistical propositions rather than a property of reality, a fact clear in . . . phrasing of discussions of causality specifically in terms of equations and conditional probability' (Abbott, 2001: 111). Abbott blames the obsession with the stern aesthetics of mathematics for introducing a conceptualization of causality that hinges on force or compulsion, while the best we may achieve in the social sciences is a contingent understanding of *association* or plausibility. Here, the configuration's parts work together to produce the outcome. Naturally, it is tempting to try to isolate the factor that was decisive in that process. That would be against the nature of configurations, however. Configurations ask the researcher to put particular pieces together into larger wholes; that is, events and event sequences need to be understood collectively in bringing forth the outcome. Following Mackie, we may identify certain factors to be insufficient but necessary parts of a condition which is itself unnecessary but sufficient for their effects (Mackie, 1980). Such conditions are configurational by nature. By definition, what is observed can only be explained retrospectively, and not predicted (Williams, 2009) – a point which we will adhere to in the analysis of our own data.

In many ways, the abandonment of variables and the focus on configurations make a researcher's life a little more challenging. No longer can we rely on the assumption that for instance search tasks for actors in search of alternatives rely on a limited set of (predefined) rules, or that such actors move within a space of limited and fixed parameters. We will now have to tease out the many types of interactions in the real world, not only in terms of quantity but above all in terms of quality, and account for all empirical manifestations. The same goes for the other building blocks of fitness landscapes, such as the many fuzzy ways in which search tasks are executed (if at all) or the changing latent and manifest preferences of actors. The only way to access such micro-motives and to connect them to macro-changes in the landscapes is by collecting longitudinal, high-resolution, in-depth data. Away from convenient abstractions, real collective decision making is a very complex affair where many actors interact in uncertain circumstances with often ambiguous outcomes. All individual actors have their own considerations, an understanding of the workings of the situation they find themselves in, and an evaluation of that situation in terms of their own particular goals. In addition, these goals may shift as the situation evolves. In short, actors develop preliminary mini-theories that guide their actions during collective decision making in complex situations (Wagenaar, 2007). Understanding collective decision making therefore means tracing those mini-theories. As argued by others (Cook and Wagenaar, 2012; Uprichard and Byrne, 2006; Wagenaar, 2007), this requires the collection of contextualized data from those actors, staying as close as possible to those actors and their peculiarities. Note how this puts us squarely in the group of naturalist approaches to collective decision making.

3.6   AN EVENT-BASED APPROACH TO FITNESS LANDSCAPES OF COLLECTIVE DECISION MAKING

Following the genealogy of evolutionary theories in the social sciences in general, and fitness landscape models in particular, we tried to get a better understanding of the science behind those models. What are its assumptions and what are the implications of those assumptions? Althusser's classification helped in explaining the ostensible inconsistencies we noticed. In his classification, preliminary testable propositions and conceptual novelties are mutually informing. As such, contradictions are not a problem but a necessary aspect of the research programme as a whole. In addition, we had to take account of the fact that fitness landscapes in the social sciences constitute a form of theory transfer from evolutionary biology. The major caveat is that social complexity is logically different from biological reality and that we have to take stock of those differences before using the fitness landscape model for our purposes. Human action and interaction, motives, and the reflexive capacities of people are the dimensions that not only set the social apart from the biological but should be the focus of the research because that is where explanation is to be found. By definition, human action and interaction are contextual. This observation drove us into the arms of critical realism – which accepts that there is a reality that can be known but does not accept the standard positivist assumption of decontextualized and general causality. The critical realist ontology gave us four focal points: synchronic emergence; time; context and configurations; and action and interaction. Putting emergence as the starting point is admittedly highly axiomatic but can be justified from the many accounts on social emergence (e.g. Elder-Vass, 2005). The other focal points can be considered inevitable consequences of synchronic emergence.

By themselves, fitness landscapes don't dictate any ontology or epistemology – a model is a tool for structuring data with a number of assumptions but not yet a full-blown perception of reality. Still, we can find ample clues in Wright's and Kauffman's works about their understanding of how reality operates. Thus it is that emergence, time, context and interaction are as central to their work as they are to our understanding of social complexity – an understanding for which we rely heavily on the works of Abbott, Bhaskar, Byrne and Ragin. This point of departure requires us to take the longitudinal view and to collect data about actors' motives to engage in certain activities, or not. For this book, we have selected a number of instances of collective decision making that we understand as emerging from a series of events or lineages, in accordance with Abbott's proposal. Human agency is the core driver of these lineages.

Since actors are likely to participate in multiple lineages, these become intertwined, and we should note the possibility of converging and diverging lineages. Indeed, some actors may be more successful in achieving their goals in one lineage through (simultaneously) operating successfully in another one. On paper, this understanding of how collective decision making takes place complies with the five prerequisites defined in the previous chapter, but we will happily concede that practice is always harder than theory, so we need to put this view to the test. In the next chapter we present our operationalized fitness landscape model with which one can structure and analyse data in the critical realist fashion.

4.   The model

4.1   MODELLING AS A RESEARCH PROCESS

Data can't transform itself magically into conclusions. A model is necessary in order to bridge the gaps between theories, data and conclusions.
This model is the centrepiece of this chapter. We do not only aim to present
the model; we also would like to encourage the reader to play with it,
extend it and use it. Therefore, each individual step from the basics to the
full model is explained in detail. We also clarify some of our considerations
regarding modelling as a process, in particular when it comes to keeping
consistency between the model and the epistemological and ontological
points of departure presented in the previous chapter. We first explain
why modelling is a good idea, even in the face of social complexity. Having
done that, we move to the actual model.

4.1.1   Why Model?

So, if we believe that social scientific research should be primarily concerned about actual and configurational or contextual interaction, why
then start modelling and putting things in neat boxes and formal terms?
Isn't it contradictory to first rebel against the simplicity of variables and
then start working with models that rely heavily on simplification and
generality and that are most often variable-based? Models as independent
objects are of limited use, indeed. As for example Cilliers (1998), Rescher
(1998) and Byrne (2002) have demonstrated, all models of complex systems
are severe simplifications of complex reality. For a model to faithfully represent a complex object, it would have to be as complex as the object it
represents; that is, it shouldn't need to be reduced in accordance with our
discussion in the previous chapter. This is clearly impossible. Models in the
social sciences, in particular those that claim to generate predictive power,
feature so many constraints and assumptions that one can be forgiven
for feeling weary about them. The question then is what a model should
represent or what its added value could be if it can never contain all the elements and relationships present in real life. That is, if something looks like
reality but is built on artificial, mechanistic premises, it does not actually

relate to reality – it is just something that bears an artificial resemblance to the real world in the same way that a picture of Thomas the Tank Engine represents an actual train.

The real utility of models is that they promote thinking and communicating about complexity (Epstein, 2008). As we argued in the previous chapter, models contribute to the advancement of the social sciences in terms of Generalities III by mapping and explicating the various dimensions of the target system. Even constructions that are considered incorrect help to further our understanding, because they force readers to think about the reasons why such a construct is incorrect and where refinement would be necessary. Building and presenting a model for scrutiny mean becoming clear about the assumptions underlying a specific approach and about the exact meaning of certain concepts. This is particularly useful in the social sciences, where concepts can (and often do) mean slightly different things in specific circumstances or in instances where the researchers are not completely certain about what they are trying to convey. Models bring, even enforce, that clarity (Klein, 2015).

Models also have an explorative and creative function in addition to bringing clarity and the ability to scrutinize certain assumptions. Models, such as the various versions of fitness landscape models discussed in Chapter 2, offer methods, techniques, assumptions and small causal chunks that invite the further exploration of a model or target system. Advantages such as these mean that the model gains its role and importance through interaction with the target system and the people working with it. The important point we are trying to make here is that the modelling process is as important as the model as a product, perhaps even more so. We are convinced that the modelling process – as a way of describing how certain properties and mechanisms of social systems may work – has given us a better understanding of evolution in collective decision-making processes. Models are

> things we can play with, as long as we recognize that playing with models is part of the process of socially constructing the social world and the intersection of the social and natural worlds. The models are, if not outside the world, then at least enough apart from it for us to do things with them which help us to grasp the world, certainly the world as it has been although there are much more profound difficulties in grasping the world it might become. (Byrne, 2002: 22)

In other words, modelling as a research exercise is very useful, as it is a way of eliciting implicit ideas, propositions and assumptions.

The five prerequisites mentioned in Chapter 2 need to govern the modelling process. Having established a clear and consistent ontology and epistemology in the previous chapter (prerequisite 1), we now need to develop

a model that is able to process data obtained from the real world (prerequisite 2), that should allow for unambiguous measurement (prerequisite 3),
that should help to uncover persistent patterns between interaction, contents and fitness (prerequisite 4), and that should enhance our understanding of collective decision making (prerequisite 5). These prerequisites can
be understood as limitations, but we rather see them as a way to stretch the
modeller's creativity. If certain options are not allowed, such as working
with simulated data, then the modeller will have to find ways around it. We
like to think that real life should be modelled for research purposes, but
not the other way around: that models show how real life should be – even
if such a point of departure sets us apart from many of the models in the
social sciences.

The demand that the model should be able to process real-world data
proved to have a large impact on the model's properties, because real-world
data is ambiguous, messy and unstructured by definition. It also became
clear that the model would have to facilitate the longitudinal nature of
collective decision-making processes. 'Time' is an important feature of
decision making, so the narrative about why certain events took place at
certain points in time, and who was involved, why and in which ways, is at
the core of understanding the decision-making process. Lineages therefore
form the basis for the reconstruction of these processes. The elegance
of possible convergence, divergence or intersecting lineages helps in being
able to deal with actors who make decisions on several issues that are one
way or another related. The fitness landscape model presented here ties
these lineages together.

4.2   THE BASICS

Collective decision making, by its very nature, concerns two or more
interacting actors. We saw such an example in Chapter 1 where the railway
operator and the Minister attempted to coordinate in order to get a good
deal or, in the case of the operator, to avoid going under altogether. But
collective decision making extends to any kind of decision making, such
as a married couple deciding to go out at night, or the husband deciding
to postpone his date with his secret and much younger lover on the same
evening so that he can go out for dinner with his unsuspecting wife (just
to be clear, we have no actual experience of such a situation). Either way,
collective decision making concerns two or more actors coordinating over
a certain issue in search of an improvement in their own situation by realizing their goal, that is, to achieve a better *fit* with the demands and pressures
from the environment. In the process, they will want to push through their

own ideas and preferences about that issue. The model we present here is primarily rooted in this issue. Attached to the issue are certain actors with their ideas and problem and solution definitions about this issue.

Of course, we're not the first analysing collective decision making, as it has been one of the core themes in the social sciences. However, we feel it is necessary to point out the basics from which we work and contrast them to some mainstream approaches. First, our analysis is an analytical reconstruction of the process. Thus we are primarily concerned with the question of *how* collective decisions are made, and much less about *which* decisions should be made. In a way, this could put us in the corner of thin rational choice models, which are unconcerned with the particular values (or goals) which individuals pursue (Hechter and Kanazawa, 1997). Thick rational choice models specify the individual's values and beliefs. The thin rational choice models are usually found in economics and social choice theory. These disciplines are interested in how aggregated outcomes can be 'rational' or 'desirable' through the individuals making rational decisions. This idea that social phenomena can be traced back to individual decisions is coined methodological individualism (Arrow, 1994; Dietz, 2000). Rational choice theory is multilevel and contains assumptions about individual cognitive capacities and values, amongst others. However, we don't build our model on the premise of individual rational actors but on configurations of actors and their interdependence, which means that we cannot build on the premise of methodological individualism.

Second, we tie decisions to actors. The term 'actor' is a very general open definition and may encompass individuals, organizations, companies, governments and the like. Of course, we are aware that if an actor is defined as an organization we overlook the intrinsic complexity of the collective decision making *within* the organization based on the diversity of decisions on the individual level of that organization. No matter what the actor size (small to big) we treat all the individual actors as if they are personalities on their own capable of making a decision related to the issue at hand (but do not regard that in terms of rationality, as explained above; see Tollefson, 2002 for an extended discussion of this argument). This is mostly for practical reasons: the longitudinal nature of evolutionary processes requires an extended time series of data, and that is always at odds with the in-depth data required to analyse the individual level, not to mention the problems of upward or downward causation (cf. Morcol, 2012).

The decision-making process about a certain issue consists of a series of connected events that together form a *lineage*. Within one lineage certain (external) events may occur that significantly impact the issue, creating the possibilities of bifurcation, entry or departure of new actors (cf. the rounds model of Teisman, 2000). In other words, the configuration of actors may change, and thus actors may change their ideas and subsequent strategies about the issue in this new configuration. A series of connected events between the event that starts and the event that alters the field significantly is structured in *fitness fields*. The lineage ends when the collective decision about that issue is *reached*. In summary, the collective decision-making process is a series of events that are all connected in one particular lineage, which is structured into one or more fields that are demarcated by significant changes, or that ends when the decision is made. The basics are shown in Figure 4.1.

Following Abbott (2001), it is up to the researcher to select and connect those events that can be plausibly connected during the reconstruction. There are many ways in which events can be connected – some ways more likely than others – so researchers will have to found their particular lineages through thick descriptions, that is, descriptions that bring together the contextual information that leads the researcher to assign an event to a particular lineage, or to leave out a particular event, and so on. It will require a considerable effort in narrating, both in words and in images, the beginnings, middles and ends and the changes of the events in the fields and the lineage. The resulting fitness field is a multidimensional and graph-theoretical space representing the dynamics within that particular field. Each dimension in the field – more on their specifics below – represents particular aspects of the decision-making process, including fitness, that allow the researcher to position actors vis-à-vis each other when coded for these particular dimensions. As several fields make up one lineage, the dynamics of these fields within one lineage will become visible by mapping the various fields into one figure displaying the movements over the fields for selected variables, that is, the movement of a specific actor as shown in Figure 4.1. However, as Abbott mentions, there can be multiple lineages that diverge, converge or intersect. In our approach this means that certain actors, pressures, decisions or outcomes in one field or lineage may intersect with another related field in another lineage.

The basic three-dimensional model can be represented visually. In this visualization, the x-axis represents N, the y-axis represents K, and fitness is represented on the z-axis. By trimming down the complications and starting off with the 'simplest' three-dimensional model we need to define N, K and fitness. We will do this in the following sections.

4.2.1 Content: Problem and Solution Definitions (PSDs)

The biological version of the model is quite clear, with N being a substantive gene with certain properties, which in our world means 'content', and K being the interaction between genes, in other words 'process'. The principal

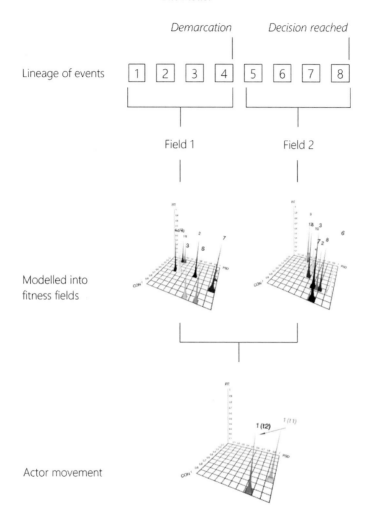

*Figure 4.1   Schematic overview of the modelling steps. The collective
decision-making process is structured in lineages of events.
These lineages are divided into fields that are demarcated by
major changes or concluded when a decision has been reached.
We then structure and visualize the state in each field in the
so-called fitness fields. These fields show the positions of all
actors involved on the grid as well as their fitness on the z-axis.
The fields are then taken together to analyse and visualize actor
movements in time. The visualization presented here shows one
actor moving between two fields (t1 to t2). Naturally, all kinds
of combinations are possible*

distinction between content and process is, in fact, a classic one in the study of collective decision making (Cohen et al., 1972), with structure being the third one. But, as we pointed out in the previous chapter, structure emerges through interaction, so we will focus on content and process. The content dimension features two idea categories that actors have about the issue. First, actors perceive and frame the issue in their own actor-specific ways, as there is rarely full consensus about the exact nature of a given issue (Fischer, F. and Forester, 1993; Rein and Schön, 1996; Schön and Rein, 1995). This means that each actor develops its own *problem definition*. Second, actors have certain specific ideas about the ways in which the issue should be solved; that is, they formulate their own specific *solution definitions*.

To denote that each actor involved in the issue has its own particular combination of problem and solution definitions regarding that particular issue, we use the abbreviation 'PSD', that is, 'problem and solution definition'. PSDs then provide us with the *N*-dimension of the model. The second prerequisite of unambiguous measurement forces us to account for all the different PSDs held by the various actors. Consequently, we need a specific scale to feature all PSDs for a certain issue in one dimension. The full problem and solution space in a complex issue is very likely to be highly dimensional, and any method would be overwhelmed by the qualitative differences between each PSD. We therefore scaled the PSDs by only looking at the *number* of problem and solution definitions for each actor while relegating their qualitative differences to the narrative. The narrowest PSDs consist of only one problem definition without a solution definition, or vice versa. The broadest PSDs actors may have are those PSDs that contain many different problem and solution definitions. When visualized on the *x*-axis, a PSD with only one problem definition or solution definition lies on the left-hand side, and PSDs consisting of many problem and solution definitions on the right-hand side. This leaves all shades of grey in between possible; depending on how many definitions are present in the PSDs they can be positioned somewhere between the two extremes. That is, all that is counted is the amount of problem and solution definitions per actor, as a proxy for substantive richness as held by the actors: a higher number of definitions can be plausibly seen as higher diversity with respect to the contents. Depending on the number of definitions in the different actor PSDs the *x*-axis ranges from one definition to the maximum number of definitions observed in a given field, that is:

$$n \text{ actors have a } PSD_i, \text{ where } \{n \in N \mid n \geq 2\} \text{ and } PSD_i \in N_0$$

We scaled the PSD-axis to fractions. As such, the axis ranges from 0 to 1, where 1 represents the maximum number of elements in a definition coded for a certain actor in that specific field, and all other PSDs are fractions of that maximum. This makes the scale dynamic and fitting to any number of definitions present in a particular field. In other words:

$$\text{every actor } i \text{ has } PSD_i = \frac{actual \text{ \# } elements \text{ } PSD_i}{maximum \text{ \# } elements \text{ } PSD}, \text{ where } PSD_i \in [0, 1]$$

The variety of different problem and solution definitions by itself is unlimited, as actors involved in a collective decision-making process on a certain issue can and will articulate their problem and solution definitions differently. This variety is of course the breeding ground for the solution, and it shows the diversity of PSDs for all actors involved. However, even though actors articulate elements of their PSDs differently, they quite often mean similar things. We will code the data for these similar elements in PSDs as being the same. For example, an actor defines the problem of a certain neighbourhood in a city as having a high crime rate, while another actor defines it as a violent neighbourhood. Although slightly different, these definitions are similar and these actors will be attributed the same problem definition. This means that the maximum amount of elements in PSDs will be lower than if these slightly different problem or solution definitions were uniquely coded.

4.2.2 Process: Connectedness

The *K*-dimension is about the interaction between genes. But, as argued before, the frequency of interaction is not very informative in the social realm because it lacks the differentiating power necessary to distinguish between qualitative different interactions. If interaction frequency fails to deliver, it is necessary to turn to an item that can approximate the relationship with PSD whilst remaining true to the vagaries of social reality, not least in order to remain true to prerequisites 2 and 3. First of all, we have to let go of the notion of independence and accept that in real-world collective decision making interaction is a consequence of the (perceived) distance, coalition forming, collaboration, beliefs, friendships and so on. A coalition may be possible because actors agree on something, or because they have had a good experience working together in the past (e.g. March, 1994), or because the main people within the coalition know each other from the golf course or share a passion for heavy metal music (cf. Hantzis, 2013). In other words, understanding interaction requires knowledge about the contents, because they are intertwined. Tempting though it may be to follow the route of independence, it is not logical and basically at odds with our thinking in terms of configurations as presented in the previous chapter. So what can we do if we accept that PSDs are tightly connected to actor interaction, to the extent that independency between the two is not very realistic, but also that the model should provide unambiguous measurement?

Actors will decide with whom to cooperate or align or, conversely, whom to avoid or even to contravene. This decision is based on actors' strategies, perceived distance between actors, and ideas about the intentions and motives of other actors (March, 1994). Thus the 'connectedness' between actors is an important driver in the nature and frequency of interaction. Connectedness is the number of actual links in a network as a rate of the number of possible links, a measure better known as *density* in social network analysis (Tichy et al., 1979) or *connectivity measure* in evolutionary network biology (Proulx et al., 2005). Thus we denote K as c_score, which defines the manner in which actors are connected to each other in the particular fitness field. This c_score shows how well connected an actor is with other actors in this issue (see also Freeman, 1978 for similar reasoning on density, and Valente and Foreman, 1979 on actor integration).

As mentioned before, events have multiple narrative antecedents. It means that we can't start from a green-field situation or a randomized situation when assessing the c_score at the beginning of a field or lineage. The actors involved can be assumed to have a certain history together that already makes them connected even without knowledge about the content of the *current* problem and solution definitions of the other actors. In other words, the starting connectedness in a field is based on (Marks, 2002; Peyton-Young, 1998; Sugden, 1986):

- actual working relations;
- the general history of the actors, that is, the ideas the actors have on each other's track record in similar situations;
- the general working history, that is, the collaboration history in previous similar situations;
- the sense of belonging, that is, mutual liking or disliking of actors.

At the introduction of the issue every actor i is connected to all the other actors to a certain degree, that is:

$$c\_score_i(t_0) = \frac{actual\ connections_i}{maximum\ \#\ connections}, \text{ where } c\_score_i \in [0, 1]$$

The c_score ranges from 0 to 1. It can be 0, as an actor can be completely new in the field, a possibility that shouldn't be ruled out but is unlikely to occur. If an actor is connected to all the other actors, the c_score will be 1.

Of course, any *c_score* between 0 and 1 is possible. The starting situation will have to be assessed by the researcher.

In the unfolding of the decision-making process the actors will inform, and be informed by, the respective PSDs of the other involved actors. Hence, actors will find out which elements of their PSDs are *similar* to elements of other actor PSDs. Similarities in the problem and solution definitions means actors will connect on content. That is, actors' connectedness with others is partly informed by the extent to which certain problem and/or solution definitions converge or diverge. Conversely, convergence or divergence in PSDs is partly informed by the perception of the position of, and interaction with, other actors; that is, actor connectedness informs the PSDs of actors. This mutual informing is the logical connection between the components in the configuration. In the configuration this means that owing to the (dis)similarity in content of elements in PSDs the link between actors and their respective PSDs will become weaker or stronger. Hence, a qualitative adjustment based on content (similar elements of the PSDs) enters, attributing weight (*w*) to the *c_score* for every actor. In other words, for every field (i.e. until a new field or solution) the change brought about by the similar elements in PSDs means that every actor *i* is now connected as follows:

$$c\_score_i(t) = w_i \times \frac{actual\ connections_i}{maximum\ \#\ connections}$$

where $w_i = \dfrac{\#\ actual\ similar\ elements\ PSD_i}{maximum\ \#\ similar\ elements\ PSD}$, and $w_i \wedge c\_score_i \in [0, 1]$

The position of actors in the field may change owing to the weight. Certain actors will be isolated, while others will have a central position, but certain actors are also more closely linked to each other than they are to the rest of the network, forming a cluster within the network (see Tichy et al., 1979 for similar reasoning).

We have now defined the two core components, namely PSD and *c_score*, as a configuration. In theory any place on the configuration can be occupied, that is, everything between the corners (PSD, CON): (0,0), (0,1), (1,0) and (1,1). Do realize, though, that in our model it is logically impossible for an actor to be involved without having any element in the PSD. That is, an actor can have only one problem or solution definition and all the others can have very many, which means the actor will be close to the corner where PSD is 0, but it will never be 0. It is theoretically possible that an actor has no connections at all with the other actors in the

field, but can have ideas about the issue, ranging from one solution definition (also known as the consultant, in the 0,0-corner) or many elements in
the PSD (also known as the well-intending active responsible citizen who
keeps being ignored by everyone else, in the 1,0-corner). Of course, the
option of being highly connected but with a limited amount of elements
in PSDs will be in the 1,0-corner. Lastly, actors that are highly connected
and have many problem and solution definitions will be positioned in the
1,1-corner. This allows mapping the actors involved in a certain issue relative to each other over time. When engaging in collective decision-making
processes, they shift in both dimensions as they opt to cooperate, or not,
with certain actors and as they try to align their PSDs, or avoid such
alignment.

4.2.3   Defining Fitness

It is plausible to assume that actors will aim to reach their goals – even
though they may be confused about what they want, or change their mind
about what they want, even if they want something entirely different from
what they expressed. It is not a far stretch to say that goal attainment can
help an actor to survive another day. However, that claim needs to be substantiated and operationalized. In its simplest form, fitness – the extent to
which a species fits a particular niche, of which multiple local optimums
can coexist in a given landscape – has a clear and dichotomous indicator:
survival or not. However, this is less clear cut in collective decision-making
processes. Lack of success in a particular issue doesn't necessarily mean
that a specific actor becomes extinct. It may withdraw from the issue or
wait for a better opportunity. Conversely, a winner can become the loser
when the boundaries of the issue shift. It can therefore be argued that
a dichotomous indicator where fitness is measured in terms of survival
(1) or disappearance (0) doesn't do justice to the complexity of collective
decision making (or, incidentally, to the complexity of biology; see Reiss,
2007). It is therefore important to tailor the concept of fitness to social
reality and the kind of issues researched in this book.

The simplest form of fitness discussed above can be seen as a form
of direct fitness, where individual behaviour favours the individual but
nothing beyond that (West et al., 2011). While such situations may occur,
it is much more likely that fitness can also be gained from helping others;
that is, fitness can be attained indirectly if one's own fitness can stay more
or less level at the same time. Since both direct and indirect fitness are very
likely to occur in social reality, it is therefore better to speak of *inclusive*
fitness, which denotes the effect of an actor's actions on its own fitness and
that of others directly related to this actor (Grafen, 2006; Hamilton, 1964),

or conversely the impact of others on the fitness of that particular actor, that is, neighbour-modulated fitness.

The closest measure of inclusive fitness is the extent to which actors reach their goals; that is, we need to think about the struggle for survival in terms of Lasswell's (1936) who gets what, when and how. Fitness can now be defined as the *probability* of an actor achieving its problem and solution definitions, that is, the actor's desired goal(s) as defined in its PSD. An actor's ability to get closer to its goal depends on its position relative to other actors, not just on its own intentions or deliberate planning but also on what others connected to that actor do. Note that we focus on *probability* for three reasons. First, goals are rarely static, and they shift as situations change, actors adjust certain elements in their PSDs accordingly and goal displacement takes place. Second, actors with highly probable goals may still never realize these owing to the movement of others and/or changing circumstances. That means that goals and thus fitness can shift in the course of time. Third, a 100 per cent probability equals actual goal achievement, which means that a decision has been reached in a field in favour of one or more of the actors, and the field is thus *concluded*.

Over time, actors will converge and diverge in their attempts to gain a better position or fitness (cf. goal alignment; Venkatraman, 1989). The probability with which actors are able to realize their PSD is expressed with a position on the z-axis: the higher the probability of realization of elements in its PSD the higher the level of fitness. In other words, a higher position indicates that an actor is closer to achieving its goals, and a lower position means that an actor is further away from achieving its goals relative to all the other actors. Inevitably, fitness values need to be attributed in hindsight when it has become clear how and which actors managed to get (elements of) their PSD realized within the specific situational context and narrative outcome. The fitness attribution is done by the researchers based on the data in each field, that is, the success probability to each actor's configuration of PSD and *c_score* in relationship to that of others in the field derived from the interpretation of the data. In modelling terms, the fitness value attributed (f) based on the narrative consequence of the events *per field* for *each actor* derives from the specific configuration of PSD and the weighted *c_score* for that actor, where v is the value attributed by the researchers:

$$f_i = v(PSD_i, c\_score_i) \text{ for each actor } i, \text{ where } f_i \in [0, 1]$$

Our model does not a priori assume a fixed relationship between a certain fitness value and a position on the grid. The question of which configuration of PSD and *c_score* leads to fitness needs to be answered empirically. We will revisit this question in Chapter 7.

4.2.4   Unit of Selection

The interaction and alignment between actors in search of fitness produce a number of evolutionary outcomes, the two most important ones being the surviving actors and surviving contents. We can obtain a rough assessment of the fitness of individual actors. However, actors rarely just disappear in the way that species can go extinct. They may give up pursuing a certain element of their PSD, pull out altogether or even change their profile, but they rarely just die out. The considerations and subsequent actions of actors act as variation and selection mechanisms for the contents. Any movement in the field means exerting selection pressures on the situation. Reciprocity means that selection pressures from one actor are met with selection pressures from other actors on the PSDs and subsequent decision-making process (Dopfer, 2005; Foster and Holzl, 2004; Gerrits, 2011). Which particular problem and solution definition comes out on top of the competition in the short and long run is therefore not a given (John, 1999). It depends on the situational feedback as well as the strategies followed by actors, for example the support they manage to amass by cooperating or going for a stand-alone strategy if they think this will bring them closer to their goals. In other words, whom actors interact with affects the results of the possible actions of actors (Arthur and Durlauf, 1997). Tracing the evolution of PSDs and the positions and actions of actors tied to those PSDs will give a thorough understanding of the mechanisms of variation, selection and retention at work.

4.3   INTRODUCING DYNAMICS

It is hard to think about evolution without thinking about time – in particular when the foundation of this model comes in the shape of series or lineages of connected fields. We will therefore discuss three types of time-sensitive dynamics: within the field (i.e. field-bound dynamics); across the different fields within the whole lineage (i.e. lineage-bound dynamics); and the coupledness between lineages, as (events in) lineages can intersect and become connected.

4.3.1   Field-Bound Dynamics

Minor movements across one field in response to selection pressures from within the configuration, but also from the environment without considerable changes in the issue, can be considered as dynamics *within the field* or field-bound dynamics. In collective decision-making processes actors seek to improve their fitness; that is, they will converge or diverge to raise the probability of realizing (elements of) their PSDs. It doesn't take much imagination to see how human actors deploy search strategies in order to find the best opportunity given their configurations and the environmental pressure. When the search leads to the identification of better alternatives than the current position, an actor can decide to move. In doing so, it may cause others to move as well. All these changes and movements are relatively minor shifts in either elements of the PSDs or c_score, because major shifts would mean a change of field.

For the modelling of the field-bound dynamics, we will have to content ourselves with the fact that some kind, any kind, of search across alternatives is deployed, and that the resulting movements of an actor are reciprocally related to the movements of others in the same field. Ideally the movements of actors within one field become visible. However, very high-resolution data is needed in order to visualize the field-bound dynamics, that is, a lot of information on PSDs and connections in many events within one field. In practice, most data is not finely grained enough to reach that goal. This role has to be taken over by the narrative of the event sequence. It will inform on how actors deploy strategies, alter their ideas in search of better fits, align to reach a higher fit, react to problem and solution definitions of other actors, and so on. One has to keep in mind that this only holds as long as there's not any major change. Events that change the configurations of the actors delineate the fields. This can be expected to happen frequently, so we will also have to look at the dynamics of the different fields within the whole lineage.

4.3.2 Lineage-Bound Dynamics

Besides the qualitative narrative to describe the field-bound dynamics, we are able to visualize the configuration and its related fitness for each field for certain time frames. Because several fields make up one lineage, the dynamics over these fields within one lineage will become visible when tying these fields together. Because our model is clearly a specific transformed version of the original model, the concept of connected fields in a lineage has no match with the existing fitness landscape literature. However, the idea that decisions (or non-decisions) that ended a certain field may impact the possibilities or options in a next field can be quite informative, for example by highlighting that decision makers often face problems created by decisions others took in the past and that seemed sensible back then; history matters (e.g. Gerrits and Marks, 2008). Evolving systems 'are sensitive to initial conditions . . . [O]nce the system "chooses" one branch over another and travels sufficiently far along that path, it stabilizes and the system settles into its new evolutionary pathway' (Reed and Harvey, 1992: 363–364). That is, the events and their unfolding in one field affect the events in a following field; the selection mechanisms in a latter field have been influenced by the decision outcomes of a previous field.

Lining up the fields in a lineage will relate the various positions of actors in the field across time. For instance, a certain actor that was able to realize most of its problem and solution definitions in a field, that is, has reached a high peak, can hardly realize any elements in its PSD in the next field, that is, a lower peak. All other kinds of dynamics can be shown. Recall from Chapter 1 that the Minister of Transport was very successful in getting a very high concession fee, as the incumbent railway operator really wanted to keep its monopoly position. Later on, the Minister had to save that same operator from bankruptcy, as it could not deliver on the concession; that is, the Minister lost a lot of money by accepting the bid that could ostensibly bring in considerable money. The movement of the actors from their aligned peaks in the first field about the concession to the lower peaks in the latter field of bankruptcy catches these dynamics in one glance.

The visualizations require one disclaimer, however. We have modelled the axes in such a way that they are relative instead of absolute. This means that the amount of PSDs and c_score in one field consist of fewer or more elements than in another field; that is, the fields' size changes – some fields more than others. The movements across the fields still need a qualitative narrative supporting the visualizations.

4.3.3   Coupledness and Coevolution

There are lineages that diverge, converge or intersect because of actions from actors or external events. Actors involved in a decision-making process on a certain issue may be involved in other related issues too. That is, an actor may be involved in two (or more) different lineages and as such consciously or accidentally connect certain elements of the PSDs in these two lineages. An example of an external event that affects two fitness fields is a national government cutting back on expenditures in a certain policy domain, and thus influencing specific fitness fields in different lineages because they affect the conditions for those lineages. Yet another example is that a decision reached on a certain issue influences the possibilities for the decision making on another issue. It may then be that the dynamics or decisions in the related issue feed back into the first issue. These feedback loops exert pressure on the issues; that is, owing to actors being involved in different but related issues, they become coupled.

These examples show that fields in two or more lineages can become coupled. This coupling of correlated fields will need to be made explicit in the decision-making processes. A somewhat similar concern is also present in biological fitness landscapes, from which we have transferred the idea of coupled landscapes. In short, this means that 'Adaptive moves by members of one species deform the fitness landscape(s) of other species' (Hordijk and Kauffman, 2005: 42). Actors start moving across the landscape as time progresses and fitness is obtained or lost within the landscape. The same dynamics will take place between landscapes; that is, changes in the relative fitness value in one landscape impact the other coupled landscapes. For example, the decision to optimize the route for 300 km/h fed forward to the future landscape when NS had to buy trains years later.

Borrowing from Hordijk and Kauffman (2005) and from the concept of the decision-making arena (Klijn and Koppenjan, 2015; Koppenjan and Klijn, 2004), we define coupledness in fields of collective decision making in two ways. First of all, there is coupledness in time, where the movements of actors, redefinition of problem solution definitions or decision outcomes in one field at t_1 affect the configuration in a field in another lineage at t_2. That is, the event at t_1 in the field of the first lineage has a delayed impact on the boundaries, possible elements in PSDs, or connections and so on at t_2 in the field of the other lineage. Naturally, this is where path dependency (Arthur, 1994; David, 1985; Pierson, 2000) comes to the fore again. For example, one can observe how certain events in the lineage of the route and track decisions of the Dutch high-speed railway had a real, tangible impact on the dynamics of lineage of the operation of the railway at a later stage when the construction had been completed (Gerrits and Marks, 2014b).

Second, we find coupledness between fields over two or more different lineages in the way that one event has an impact on two fitness fields simultaneously. That is, owing to an internal or external event, rules, conditions, PSDs, connections and so on are influenced at the same moment. If the movements *within* the field constitute adaptation of actors, the simultaneous reciprocal influence *between* fields in different lineages constitutes coevolution (e.g. Gerrits, 2008; Kallis and Norgaard, 2010; Moreno-Penaranda and Kallis, 2010; Norgaard, 1984, 1994, 1995 on coevolution in decision making). An internal event may be that the movements, PSDs or decisions of actors in one field at t_1 have an impact on the PSDs and fitness values in another related field simultaneously. Such couplings have also been noted in game theory (Henry et al., 2011; Knott et al., 2003; Lavertu and Moynihan, 2013). Again, the high-speed railway case can serve as a good example here, as negotiations between the Dutch and Belgian governments over the exact alignment of the railway across the border had a reciprocal influence on the negotiations over the deepening of the Dutch Westerschelde estuary to extend the Belgian port of Antwerp, as will be more extensively demonstrated in the next chapter. An external event may be a decision, rule and so on of an unrelated actor that redefines the range of possibilities of the PSDs for the actors in two lineages simultaneously.

In other words, in our interpretation there are two versions of coupledness: asynchronous coupledness, where an event in one lineage affects the configuration in another lineage at a later moment in time; and synchronous coupledness or coevolution, where one event affects two lineages simultaneously. The coevolutionary relation can be depicted by borrowing the interaction classification of Odum (2004). Odum classifies the interaction between species in different forms of coevolution; for example, a species may benefit from the interaction with another species, while that other species is not affected. Interaction can be beneficial or detrimental, or all shades of grey in between, for either or both populations of species.

As mentioned above, the unit of selection is formed by the elements in the respective PSDs. We can assess the coevolutionary relationship by looking at the changes taking place with regard to the PSDs. For example, if the coupling means that actors are getting more connected in a certain field, it means that they utilize this to fulfil their PSDs or that they redefine elements of their PSDs. It is possible to classify the coupledness of any amount of lineages, but for the sake of simplicity we present the classification for two coupled lineages (see Table 4.1). Table 4.1 provides the theoretical classification for all possible couplings. This table departs from Odum in one way: we leave out the theoretical option that coevolution may not have an effect on one of the lineages involved. Cases where one lineage is not affected (i.e. a 'zero' value in Odum's classification) constitutes a unidirectional relationship between an event in one lineage and the configuration in another lineage. This is almost similar

Table 4.1   Classification of coupledness

Scaled type of interaction	L1	L2	Nature of coupledness
0	−	−	Very unfavourable
	0	−	Somewhat unfavourable
	−	+	Mixed results
	+	−	Mixed results
	0	+	Somewhat favourable
10	+	+	Very favourable

to asynchronous coupledness, but then simultaneous in two lineages. In other words, the 0 in Table 4.1 constitutes a form of coupling but not coevolution. A complete neutral coupling (0, 0) would empirically be redundant, as this would be coupling between lineages where no mutual influence exists.

4.4   FINAL THOUGHTS

The purpose of this chapter was to introduce the fitness landscape model central to this book. In the process, we converted N to problem and solution definitions and K to connectedness between actors. An important modification here is that we consider NK to be a configuration; that is, K bears with it both content and process, in particular in relationship to past experience with other actors in similar and the same cases. We then introduced the third dimension, namely fitness, which we defined as the probability of an actor achieving (elements of) its PSD as a result of the adaptive moves in relation to the adaptive moves of others. We opted for relative fitness values, as they allow us to mark the success rate of particular actors as higher than, the same as or lower than those of other actors. These three dimensions gave us our basic model for collective decision making. To this, we added PSD as the unit of selection that produces telltale signs of the evolutionary process.

The point where things really get interesting is when time is introduced. We focused on three aspects here. First, we presented the field-bound dynamics as a tangible marker for time. Field-bound dynamics are started as actors try to gain and maintain higher individual fitness in response to the moves of others that alter the average fitness in the field, thus requiring such a move. Second, we presented lineage-bound dynamics as events and their unfolding in one field affecting the events in a following field. Third, we presented coupledness between fields in different lineages where movement or changing environmental conditions in one field have an impact on other coupled fields.

We now have all the ingredients in place for a complete fitness landscape model for collective decision making. The reader may notice, as many authors have, that such a model allows the exploration of many dimensions of evolutionary processes. Exploring them requires that the model is put to the test. We do this in the following chapters. Chapter 5 features the baffling story of the attempts to build a high-speed railway in the Netherlands – and subsequent failure of that attempt. We explore and test all the basic aspects of the model in this highly detailed narrative. Chapter 6 takes a closer look at the three dynamic aspects above. In order to keep the narrative clean, we have placed all the methodological details in the appendices. For each study in Chapters 5 and 6 we gathered empirical data. The overview of the data and how we collected and processed it can be found in Appendix B. The actual coded data for the HSL-Zuid study can be found in Appendix C.

5. Memory of a dream: high-speed rail in the Netherlands

5.1 SETTING THE COURSE

By the time you read this, the Dutch high-speed train sets mentioned in Chapter 1 will all have been returned to their manufacturer in Italy, which is going to try to sell them somewhere else. Meanwhile, Netherlands Railways (NS) is working with a makeshift replacement service for the cancelled high-speed connections until a more permanent solution has been developed. It marks a temporal conclusion of a long-running saga going back all the way to the 1970s. We can be certain that there will be new episodes in the future, as Netherlands Railways is hell-bent on keeping the high-speed concession running and issued a tender for new train sets in early 2016. The decades-long attempts to develop and operate a high-speed railway service in the Netherlands serve as an example par excellence of the evolutionary nature of collective decision-making processes. The sheer length of the process means that no single individual was involved from start to end, and the very real massive costs sunk into the construction, operation and maintenance meant that the process became entrenched and difficult to abandon (cf. Gerrits and Marks, 2014b). Consequently, ministers, civil servants, planners, builders, financers and many others plodded on, in the process creating a lumbering beast that was very hard to control.

The idea of building a high-speed railway link didn't come out of nowhere, of course. It was the Japanese Shinkansen experiment in the 1960s that inspired the French to develop and test their own high-speed train, or *train à grande vitesse* (TGV), in 1972. In turn, this sparked discussion among Dutch politicians, civil servants and Netherlands Railways about having high-speed trains too. Thus it is that a first proposal was made in 1973 to connect Amsterdam with Belgium via Rotterdam, the so-called AmRoBel plan. The idea was extended with proposals to improve national and international rail traffic in an attempt to offer a viable alternative to long-distance road traffic and flying – in particular for flights of 500 to 1000 kilometres.

Years of political debate ensued. The formal planning procedure was started in 1987 and encompassed three major planning studies centring on the creation of a so-called corridor running from Amsterdam in a southerly direction to Rotterdam and Brussels. This was to be called the Hoge-Snelheidslijn Zuid (high-speed railway link south, or HSL-Zuid). The studies encompassed a formal feasibility study, the environmental impact assessment of all possible variants, and the route decision. There were multiple alternatives for track alignment, but from the outset it became clear that the Minister for Transport was strongly in favour of a new and dedicated track instead of upgrading existing tracks. In addition, the new track had to be as short as possible and to allow a maximum speed of 300 km/h. As we have argued elsewhere (Gerrits and Marks, 2014b), these preferences had a defining and path-dependent impact on the scope of the feasibility studies and the project as a whole.

When the proposal was sent to Parliament for approval in 1991, it met severe criticism from both its members and the media. First, the costs were estimated at €1.45 billion (i.e. fl. 3.2 billion; hereafter we will use the euro (€) for consistency), of which only €680 million was actually covered in the national budget. Second, it was widely – and rightly, as we will see later on – believed that the budget estimate was too optimistic and that a budget of €2.25 billion would be more realistic. Third, and in response to the possible budget deficits, the Minister announced that 50 per cent of the construction costs would be paid for by the private sector in a public–private partnership arrangement. However, the private sector expressed little enthusiasm for such a deal. Fourth, the choice of a new and dedicated track was considered poorly motivated, especially given the marginally shorter travel times in comparison to alternatives. Fifth, lack of agreement with the Belgians about alignment across the border meant that the plan was not as robust as presented. If the Belgians disagreed, the whole alignment would have to change. Given these considerations, it is no surprise that the whole proposal was rejected and that the government had to return to the drawing board.

A second and major revised proposal was submitted to Parliament and approved in 1994. When the plans were presented, the government kept pushing its preference for a new track, which was now called the A1 route from Amsterdam to the border with Belgium (see Figure 5.1 for a map with the final alignment). A1 was the shortest possible route with the highest average speed but also the most expensive one because of the costs of land acquisition and the construction of a considerable number of tunnels and viaducts (Algemene Rekenkamer in Tweede Kamer der Staten-Generaal, 2003). This preference was pushed through Parliament, which gave the go-ahead for the plan. Since that moment, 20 years of decision making about designing, planning, building and operating the HSL-Zuid have resulted in a project that spiralled out of control and that cost €7.1 billion or about

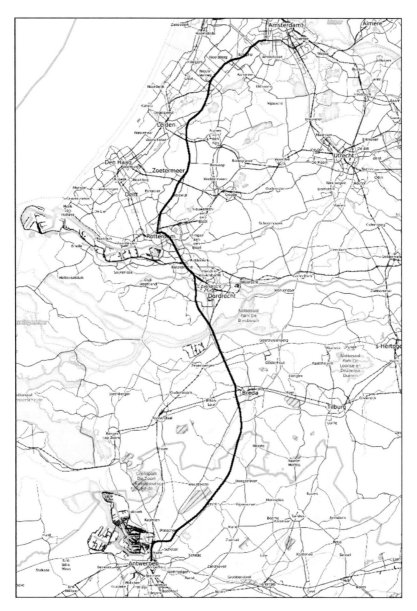

Source: Custom map courtesy of inkatlas.com and OpenStreetMap.

Figure 5.1   *Map of the HSL-Zuid study. The HSL-Zuid runs from*
               Amsterdam Schiphol Airport in the north to Antwerp in the
               south, via Rotterdam (approximately in the middle of the map)

three times more than estimated. Other outcomes included an operator that had to be saved from bankruptcy three times in a row, a severe lack of proper services covering important parts of the Dutch network, and a train manufacturer on the brink of collapse.

While some of the problems with this project can be attributed to common planning mistakes, we will argue that the mutual movements of all the actors involved over time created a rather erratic arena where an actor could win one round, only to find itself the loser in the next round. Using the fitness field model, we will map the movements of actors to demonstrate, analyse and visualize the building, tendering and operation of the HSL-Zuid from the decision in 1994 until the collapse in 2013. We divided the two decades into four lineages. After the decision to build the shortest possible route, that is, from Schiphol to Rotterdam in a fairly straight line through the so-called 'Green Heart' of the region, the route needed to be financed and built (lineage 1). Simultaneously, there was an ongoing debate and series of negotiations about the connection across the border with Belgium (lineage 2). Once the track was built, decisions about the concession, that is, who would obtain the right to operate it, needed to be made (lineage 3). The winning operator would be confronted with remnants from past decisions and new complex situations – such as trains that weren't fit to run – while trying to operate the HSL-Zuid concession (lineage 4). We'll start our discussion in 1995. As we have said before, a high level of detail is necessary to understand what has happened and why it has happened. In other words, this will be the beefiest chapter of the book! We are aware that some will wonder if such a detailed description is really necessary and why we don't just focus on the core findings. The short answer is that, for once, we want to show how one moves from raw data to conclusions. We will present the narrative in the present tense in order to show how the decision-making processes unfolded and how actors – who were obviously and blissfully unaware of what would happen later on – tried to deal with the situations they were facing.

5.2   LINEAGE 1

5.2.1   Decisions about Financing the Construction of HSL-Zuid

Lineage 1 consists of an extensive antecedent narrative, one field and a narrative outcome of the decisions in each of the fields. The antecedent concerns events about how to share financial risks in building the HSL-Zuid, but no real decision making occurs. In the only field in this lineage,

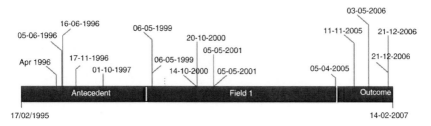

Figure 5.2   Overview of events, lineage 1

decisions are made about the financial consequences of the tendering process. The different events in the lineage are visualized in Figure 5.2.

Antecedent – going private (17 February 1995–20 January 1999)
After the decision to build the HSL-Zuid the Cabinet has to find the finances to build it. A considerable budget comes from the funds generated through the selling of natural gas and state companies, that is, €2.25 billion. This money will be divided over several larger infrastructural projects, and it is not clear how much will be reserved for the HSL-Zuid. It is clear that the reservation will certainly not be enough to cover all expenses for the HSL-Zuid. This becomes even clearer as during the decision about the route and track (see lineage 2, field 1) an additional amount of about €1.1 billion is needed to build the Green Heart Tunnel and to compensate villages along the route. These decisions were necessary to preserve a relatively pristine natural area, raising the total building costs to an estimated €3.5 billion.

This period is marked by a belief that the market can solve problems with public service delivery. The belief is driven by the waves of privatization in the UK in the 1970s, including deregulation of the railway sector and subsequent abolishment of British Railways. Privatization also took off in the Netherlands at the end of the 1980s and meant that new forms of financing were now deemed viable, that is, tendering of tasks and public–private partnerships in projects. Congruent with the discovery of the possibilities of private market mechanisms, the Minister of Transport aimed at a public–private partnership with construction and financial companies to share the costs and benefits of the project. The Minister expects €600 million to be financed by private parties. To get there, she organizes a dinner for private companies to get them interested in financing, building or operating parts of the HSL-Zuid either completely or in some form of public–private collaboration. Companies such as Siemens, HBG and Générale des Eaux are showing an interest, but major private investors like

ABP and ING are reticent. The Dutch rail operator NS is not interested in building, but is interested in operating the HSL-Zuid (see lineage 3 for further details on the operation of HSL-Zuid and the role of NS).

Meanwhile the Minister also makes clear that a 50 per cent cost overrun for the HSL-Zuid is acceptable. This means that in the event that there are cost overruns these should be financed by windfall profits in other infrastructural projects. One of the cost overruns already surfacing is that many minor adjustments to the plan are made by the Minister of Transport, costing €340 million extra.

After the aforementioned first attempts to attract the interest of private companies the Minister of Transport sends a consultation document to different companies to see whether they are still interested and, more importantly, how their knowledge can be utilized to improve the HSL-Zuid building plan. More than 100 companies, mainly construction companies, are interested in participating in the HSL-Zuid; that is, they are mainly interested in building train stations or the exploitation of the super-fast trains. After these attempts to get parties interested in financing, building or operating parts of the HSL-Zuid either completely or in some form of public–private collaboration, the Minister decides that the tracks, catenary, noise barriers and signalling systems will be put to tender in design, build, finance, maintain and operate (DBFMO) contracts. The estimated costs for building and maintaining rails and catenary are between €450 million and €700 million. The building companies will also be responsible for maintenance for the 30 years after construction, for which another €700 million will be needed. Owing to the high risks, the government will build the foundations, tunnels and shell constructions itself through standard design and construct contracts, at estimated costs of €2.7 billion. The Minister thinks that the costs will be compensated by the income generated through the concession for operating HSL-Zuid.

Field 1 – building can start, but fraud is detected (11 February 1999–5 April 2005)

The House of Representatives is becoming a bit anxious about the private financing plans of Minister Netelenbos, as it is all over the news that several recently concluded building projects have had major budget overruns as a result of 'bad' governmental contracts with builders and financers. The House wants more explanation and details of the building and financing plans from the Minister. In the clarification by the Minister it is pointed out that tendering for building the foundation is split into six separate tenders. Each tender for one segment of the building of the foundation will cost approximately €450 million. Even though it is split into six segments the costs will still be so high that only major national and international firms are able to participate. The Minister also announces that four building consortia are contending to build the rails, catenary and signalling systems up to the Belgian border. These are three consortia headed by major Dutch builders, that is, NBM Amstelland (partners are Siemens and Deutsche Bank), Ballast Nedam (partners are Daimler-Benz and Arcadis) and VolkerWessels Stevin (partners are Strukton and ABN AMRO), and one from the USA, namely Bechtel. The Minister of Transport starts a new tender for the section between the north part of the tunnel and New Vennep (the northern part of the route), as the offers already received are deemed below par.

Just a limited number of builders are capable of and certified for building high-speed rail infrastructure. They are therefore able to form a strong negotiation position. This active builders' market, together with rising land prices, increases the cost for the HSL-Zuid by another €100 million. Ultimately the Minister of Transport makes deals with five different consortia that will build the foundations on different parts of the route:

- Brabant South: German Philip Holzmann with Dutch Heijmans, NBM, HWZ and HAM;
- Brabant North: Dutch VolkerWessels Stevin, Ballast Nedam, Strukton, Vermeer and Boskalis;
- Green Heart–Rotterdam: Dutch NBM Amstelland, Heijmans, HBW (HBG group);
- Routes Schiphol–Green Heart and Rotterdam–Moerdijk: Dutch Ballast Nedam, Van Hattum en Blankevoort and Strukton (a former NS company);
- Tunnel through Green Heart: French Bouygues and Dutch Koop Tjuchem (see also lineage 2, field 2).

The Minister signs the design and construct contracts with the five consortia at an estimated cost of almost €1.9 billion, which is significantly higher than the initial planning. The House of Representatives is not amused by the cost overrun, but is even more astonished to find out that the Minister has granted the tender for parts of the construction of the foundation to two companies that are under suspicion of having committed fraud, that is, HBG and Strukton. These two companies are under investigation by the prosecution, as they are believed to have overcharged the Minister through falsified invoices for building the Schiphol Tunnel.

Meanwhile the costs for building the HSL-Zuid have gone up again. This increase relates to €250 million of expected extra costs for the catenary and other technical equipment, as well as an unexpected €350 million for the foundation. However, the building process and contracting have to continue, so the Minister announces that she will sign a contract with the newly formed Infraspeed consortium for €2.5 billion. This consortium was one of the two selected consortia that were allowed to compete for the contract to design and build the rails, safety systems, energy supply, technical installations, noise barriers, maintenance and finance (i.e. DBFMO contract) for the 25 years after the building process. It also becomes clear that building will be delayed again, as there are difficulties in acquiring the necessary land.

In the middle of the summer of 2001 it becomes clear that there is some disagreement between the House of Representatives and the Minister. The House is of the opinion that the Minister should inform the House better about the continuous cost overruns when executing major infrastructural projects. An independent temporal committee chaired by MP Adri Duivesteijn investigates the decision making and control on large infrastructural projects in the Netherlands to come up with improvement suggestions. A couple of years later the committee reveals that the Minister of Transport has kept the risks of the Betuweroute (dedicated freight railway line) and the HSL-Zuid hidden from the House of Representatives for years. It also shows that there were many consultations between the building companies about how to deal with the information provided by the government about the tendering process. The builders confirm that they had split themselves into four consortia, with one of the builders as main contractor and the others as subcontractors. The four consortia each submitted to only one of the parts of the line. This way they had all the information for the whole project, and thus they were able to delay agreement between government and builders and raise the price by about €800 million. These price agreements between the building companies may be a reason for the government to claim compensation from the builders. The Authority for Consumers and Markets (ACM) ascertains that the six largest builders had made agreements on how to divide the tenders and related work for the HSL Zuid. Thus it fines the builders that had made price agreements over the last few years; the fines range from €2000 to €19 million for the biggest offender. It is not made public which firms in the HSL tender are being fined for what amount. Owing to these fines and the questions about who has to pay for what and who is to take part, the playing field is changed completely.

Outcome – building concessions are settled, but there are still some financial incidents (13 July 2005–14 February 2007)
After penalizing the fraudulent builders no further major financial decisions concerning the building of the project are undertaken. Of course, many small incidents that cost money pop up, but these are managed by

the Minister and therefore there is no collective decision making. The aftermath of the decisions about financing the construction is as follows.

The Minister of Transport will request independent research by the Court of Audit into the (financial) setbacks of the HSL-Zuid. One of those financial setbacks is that the construction of the Green Heart Tunnel will cost €6 million extra for completion, and the builder Infraspeed will have to make up for the delay. Also starting in 2007, the government will have to pay €120 million per year to Infraspeed for the maintenance of HSL-Zuid even though no trains are in operation yet. Another setback is that there are certain problems with the safety system, the European Rail Traffic Management System (ERTMS). The Minister and Siemens blame each other for the ERTMS delay. The Minister accuses Siemens of not assigning enough engineers and requesting an extra €30 million budget. Siemens, meanwhile, claims that the Minister waited a year with her decision on the tender, which caused the delay. The end result is prolonged negotiations about the tender before the building of ERTMS actually started (see also lineage 2, field 3 and lineage 4, field 2).

Notwithstanding the announced and more or less expected major cost overruns it turns out that the financial setbacks have been underestimated; they are almost double. One of the reasons is that the bookkeeping of HSL-Zuid has been done so badly that certain setbacks have been overlooked. Some of the political parties are so agitated that they request the Minister to resign, but she escapes a vote of no confidence as it is her last day in office anyway and a new coalition will be formed after the election. However, she does acknowledge that mistakes have been made in the decision process of building HSL-Zuid and that the financial consequences of all the troubles surrounding the implementation of ERTMS have been underestimated. One of the reasons, she says, is that the parties involved in building the HSL are not exchanging enough information as they have grown bitter towards each other.

5.2.2 Fitness Fields of Decisions about Financing the Construction of HSL-Zuid

Based on the narrative the connected events and fields in this lineage look as follows. For overviews of this and also the next lineage, the black bar in the middle covers the whole lineage (as in Figure 5.3). The grey bars are the respective narrative antecedent, field and narrative outcome of the lineage. In these slightly larger bars the different events with their respective problem and solution definitions (PSDs) are presented.

For the rest of the study all the actors are attributed numbers, which will be consistent through all lineages. The actors are numbered as follows:

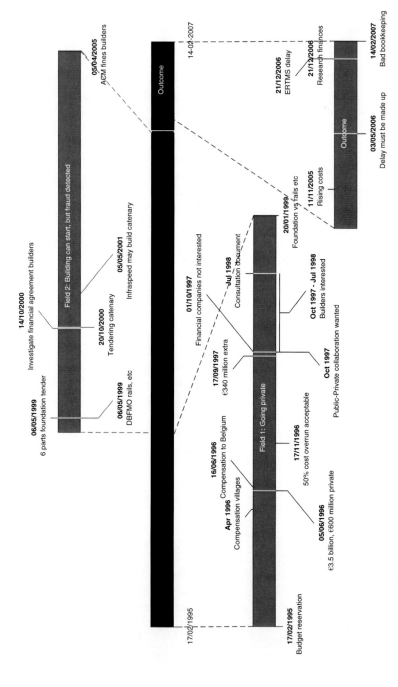

Figure 5.3 Extended lineage 1

1	Alstom	10	House of Representatives
2	AnsaldoBreda	11	Minister of Transport
3	Ballast Nedam	12	NBM Amstelland
4	Belgian Minister of Transport	13	NMBS
5	Bombardier	14	NS (HSA)
6	Bouygues/Koop	15	Siemens
7	Connexxion	16	Villages South Holland
8	DB and Arriva	17	VolkerWessels
9	Holzmann		

In this first lineage only seven out of these 17 actors are present, and they have fitness attributions based on the following argumentation: The House of Representatives wants an investigation into the contracts with the builders, which is completed by the Duivesteijn committee and detects the fraud; that is, the House has a fitness value of 1. All the builders (actors 1, 2, 5, 8 and 9) are attributed a fitness of 0.7, as most of them got what they wanted, but owing to the fines for fraud they did not get it fully. For the Minister of Transport the fitness is 0.5, as she got contracts with the preferred actors, but at much higher costs than needed due to the clustering of the builders (Figure 5.4).

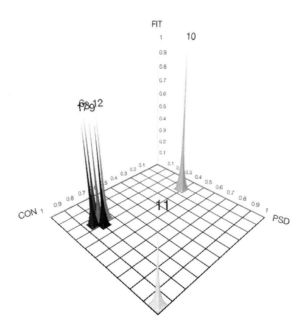

Figure 5.4   Lineage 1, field 1

5.2.3   Observations Regarding Lineage 1

Lineage 1 is a relatively simple lineage, as much of the actual decision making is done by one actor in response to emergencies such as the fraud and the problems with ERTMS. Only the collective decision-making process of the tendering for building parts of the HSL-Zuid is represented in a field. There are also not many actors involved in this decision-making process, and not much movement. This all very clearly shows the structure of a tendering process. The builders cluster and hold their position easily as a result of lack of competition and owing to mutual coordination. The Minister on the other hand is isolated as a demanding party, with little room to manoeuvre owing to the fact that she has already gone over budget and because the construction companies have teamed up.

Of course, as there is only one field, not much can be said about the field dynamics. The only thing we can attribute right now is that the starting conditions have an impact in later fields, for example expenses starting to mount up. However, this impact is not yet as great as it will be later on, as non-ergodicity will reveal itself in the long run (cf. Gerrits and Marks, 2014b).

5.3   LINEAGE 2

5.3.1   Route and Track Decisions on HSL-Zuid Infrastructure

Almost simultaneously with the first decisions on how to finance the building of HSL-Zuid, decisions about where the track should cross the border with Belgium, how to go through the pristine Green Heart of the Netherlands, and safety issues need to be decided on. The events that make up this lineage of route and track decisions is depicted in Figure 5.5.

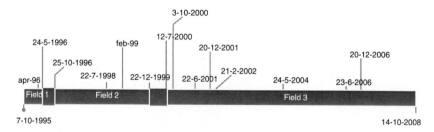

Figure 5.5   Overview of events, lineage 2

Field 1 – crossing borders and a tunnel (7 October 1995–24 May 1996)
The Dutch and Belgian governments are debating where the HSL-Zuid will cross the border, even though earlier that year an agreement was signed. In this initial agreement, the Belgian Minister Van den Brande agrees with the location of the border crossing. In exchange the Dutch government agrees that the Westerschelde, which is in Dutch territory, will be deepened in order for ships to reach the harbour at Antwerp. They are still debating because for the Dutch it has been calculated that the route near Breda is about €150 million cheaper compared to the Roosendaal alternative, while this is exactly the opposite for the Belgians. The Dutch government is trying to persuade the Belgian government to opt for the Breda route by offering to compensate the Belgians for a part of the extra cost.

Meanwhile several villages along the route are complaining that the track will cause a nuisance for their citizens. Thus several minor local adjustments are made. A decision is also made that a tunnel is the best option for the route through the Green Heart between Schiphol and Rotterdam. Of course it is open to debate what kind of tunnel and how long it should be, but for now environmental organizations, the House of Representatives and the local municipalities in the Green Heart support the idea of a Green Heart Tunnel. After about a year of quarrelling the Dutch and Belgian governments reach an agreement that the HSL-Zuid will cross the border parallel to highway E19, which is the option preferred by the Dutch government, but it will compensate some of the extra costs incurred by the Belgians.

Field 2 – versions of the Green Heart Tunnel (25 October 1996–22 December 1999)
Notwithstanding the support for a tunnel though the Green Heart the House of Representatives is not certain about what kind of tunnel should be built. Their uncertainty shows itself quite clearly, as the Minister has to answer 468 questions about the Green Heart Tunnel. She clarifies the trade-off between economic, environmental and aesthetic values, and that a decision has to be made on a tunnel length of 6 or 9 kilometres. Just as it seems that these answers are helping in reaching a decision, many local municipalities in the Green Heart, but also environmental organizations, are openly questioning the necessity of high-speed trains going through the Green Heart. This societal unrest together with the stated financial risks (see lineage 1) causes several political parties in the Senate to express their doubts about the necessity of the HSL-Zuid, while others are still supporting the high-speed rail link. Thus the Senate is divided on the spatial planning decision (PKB, in Dutch) for HSL-Zuid. However, the

Cabinet perseveres with the plan of building a track suitable for high-speed trains, as they think this is necessary for travel time reduction.

There are two more suitable options for building the Green Heart Tunnel, that is, cut and cover or drilling. Several political parties in the House of Representatives prefer the cheaper cut and cover option, but the Minister opposes it, as this would mean adjusting the spatial planning decision and she wants the builders to gain experience by drilling in unstable soil. The political parties drop their demand for the cheaper option on the condition that the Minister will do her utmost to keep the cost for the drilled tunnel to a minimum. Now that it is clear what technique will be used for the tunnel it is still under debate what the length of it should be. There are two options on the table: a tunnel of the original length of 6.4 kilometres or a cheaper version of 2.6 kilometres. The many villages along the route clearly promote the longer version, as they fear a nuisance from a shorter tunnel. The Minister starts the tendering process for the tunnel. Builders can submit their design as long as they meet the specific requirements for the tunnel, that is, safety and comfort for passengers, trains being able to go through the tunnel at 300 km/h, and two separate tubes in the tunnel or in the event of one tube the pressure wave not being a nuisance for passengers. Even though a partial adjustment of the spatial planning decision has been submitted for the alternative shorter tunnel, the Minister of Transport publishes a report that shows that the shorter tunnel will cause much more damage to nature. It is unclear how many plans have been submitted, but eventually just before the new millennium it is announced that the French/Dutch combination Bouygues/Koop will build the new Green Heart Tunnel for €430 million. The 6.4 kilometre tunnel will be a single tube with a diameter of almost 15 metres, except in the middle of the tunnel, which will hold a small section of two tubes. The 15 metre diameter is necessary to prevent pain in passengers' ears caused by air pressure.

Field 3 – building and safety (13 July 2000–14 October 2008)
Now that it is clear how the Green Heart will be spared the House of Representatives is surprised about the crossing of the railway with highway A4. The plan is to build a viaduct, nicknamed 'pergola', that is practically connected to the Green Heart Tunnel. According to the House of Representatives the concrete-pillared pergola contradicts the preservation of nature, which was the most important reason for building the Green Heart Tunnel in the first place. The Minister considers that the proposal for a heightened viaduct can be abandoned only if the province and municipalities are willing to contribute financially to realize a more expensive lower crossing.

This lowering of the tracks is a recurring issue elsewhere. In Bergschenhoek, the tracks will have to be lowered in order to reduce nuisance for citizens. Of course, this adjustment is not without extra costs, namely €20 million, and it also means that the building of the track will take about half a year longer. Later on the builders are granted nine months extra to build, as the change from a viaduct to a deepened track also means extra testing for safety. It is not the only reason for delay; the track building will take almost a year longer as a result of long-lasting negotiations with building consortia, problems with acquiring the necessary ground and several minor adjustments to the track. For instance, the contract between Bouygues/Koop Tjuchem and the government is adjusted because several revisions, including new safety measures, have to be made while building the tunnel.

The House of Representatives is threatening to stop the building process, as polluting heavy metals (\m/) have been found in the water near Breda as a result of the building of the HSL-Zuid. It wants to know exactly what the contamination risks are because of the used furnace slacks and even worse the possibility of fly ash, that is, particles of ash that may contain incompletely burned fuel and other pollutants. The Ministers of Transport and of Environment are aware of the risks of using furnace slacks, but the usage is within the building materials decree. Even though it is within the decree it turns out that the slacks have been positioned too close to the water surface and have to be removed from the construction site, which means yet another delay.

After a quiet period with no (major) incidents or adjustments to the HSL-Zuid building process the discussion starts again in 2004. The Minister of Transport announces that ERTMS will be the preferred safety system for the HSL-Zuid, even though the system is not fully developed yet (see also lineage 1 outcome, and lineage 4, fields 2 and 3). However, she is convinced it will be fully developed by the time the HSL-Zuid is finished, because ERTMS will have to be built into many European tracks by that time, owing to European regulation.

Even though the last piece of rail has been put in place the complete delivery of the HSL-Zuid is delayed once more, as the implementation and testing of ERTMS will require much more time. New safety and steering elements need to be added to the system that comply with new European norms, and also need to be in congruence with the safety systems of the existing track. The system will be upgraded and tested once more. Although the tracks haven't been brought into use yet, the first repairs and new measures need to be carried out to handle the substantial subsidence of 800 metres of track and foundations near Rijpwetering.

Except for minor incidents, for example frost damage, birds flying against the noise barriers, dust accumulation in the tunnel from trains going through at high speed (it later turned out that the fire-retardant layer on the walls and ceiling had come off as a result of the airflows), noise barriers that have trouble withstanding the pressure waves of high-speed trains passing, and so on, building continues, and on 14 October 2008 the HSL-Zuid is ready for operation. Trains are able to travel from Amsterdam to Brussels (212 kilometres) in 1 hour and 46 minutes, with an average speed of 120 km/h including stops:

- In the Netherlands, the track length is 125 kilometres, of which 85 kilometres are high-speed rail.
- In Belgium, the track length is 87 kilometres, of which 35 kilometres are high-speed rail. The remaining 52 kilometres have an allowed maximum speed of 160 km/h, with exceptions at Brussels (50 km/h), in Mechelen station (100 km/h) and in the tunnel in Antwerp (90 km/h).

Of course, in the years that follow incidents may occur which will be part of maintenance, but the building process is finished. This concludes the building and safety process of HSL-Zuid.

5.3.2   Fitness Fields of Route and Track Decisions for HSL-Zuid

The narrative of this second lineage, that is, all the connected events and fields, is shown as an overview in Figure 5.6.

The fitness attribution for the respective actors in the figures below is based on the following argumentation:

In the first field the House of Representatives is the only actor fully getting what it set out to reach – a tunnel through the Green Heart, that is, a fitness of 1. The Belgian Minister of Transport gets what he wants, but at higher cost, that is, a fitness value of 0.7. This same line of reasoning applies to the Dutch Minister of Transport, who gets the crossing at the location she wants, and who wants a tunnel, but at higher cost, that is, a fitness value of 0.7 (Figure 5.7).

In the second field the Villages of South Holland get what they want – the longer tunnel, that is, a fitness of 1. The Minister of Transport completely gets what she wants, that is, a fitness value of 1. However, the House of Representatives does not get its preferred cut and cover tunnel, that is, a fitness of 0 (this makes it hard to see in the field, but it is positioned just behind the fitness peak of the Villages of South Holland) (Figure 5.8).

In the last field the House of Representatives gets one out of two of its solution definitions, that is, a fitness value of 0.5. The Minister of

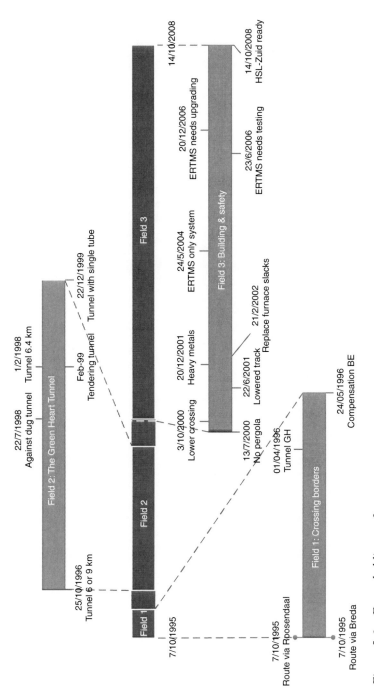

Figure 5.6   Extended lineage 2

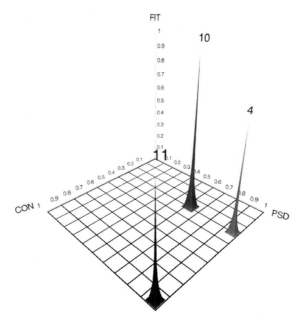

Figure 5.7   Lineage 2, field 1

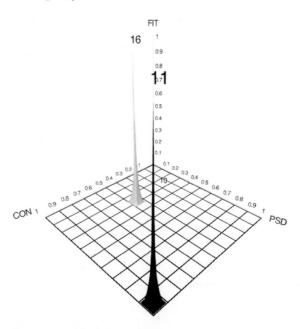

Figure 5.8   Lineage 2, field 2

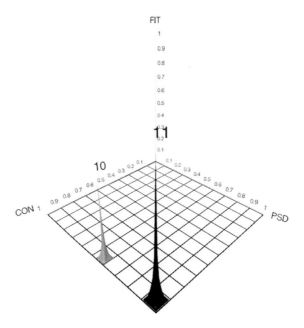

Figure 5.9 Lineage 2, field 3

Transport gets what she wants but needs to pay for all sorts of delays and minor building incidents, that is, a fitness value of 0.8 (Figure 5.9).

5.3.3 Observations Regarding Lineage 2

As in the first lineage there are not many actors involved in the decision-making process. Hence few dynamics can be observed. It does make sense that there are not many actors, as it is the Minister of Transport who needs to decide where to cross the border with only one other party, that is, the Belgian Minister of Transport. The same argument holds for the decision where and how to cross the pristine natural area of the Green Heart and access to and from it; here the Minister needs to decide with feedback, objections and so on from opposing villages or undecided members of the House of Representatives. This clear central role for the national deciding actor is plainly visible in Figures 5.7 to 5.9, as the Minister of Transport is constantly presented in the bottom right corner; that is, the Minister has the most connections and the most elements in the problem and solution definitions. What isn't directly visible is that, although the Minister holds this central position, she holds it in different configurations. In the first field the Minister is the central actor with Belgium, while she is the central actor in the tunnel debate with the House of Representatives and the villages in the second and third fields. It is a known fact that central national government ministers are linking pins in large international infrastructural projects, and this is made visible through the model. This actor is operating at multiple levels at the same time.

The position of the House of Representatives is clearly a different one. It jumps around as the MPs are (constantly) reacting to the Minister or societal pressure that shows itself through the press. The political parties want to make a good impression by showing their involvement to the public by reacting to these events, but these serve the purpose of possibly getting re-elected or other political ends. However, one has to keep in mind that the House of Representatives is a bundle of different political parties in the Netherlands. That is, the number of political parties ranges from 8 to 14 or more depending on the period in time. To see how the different political parties behave and react to events or the Minister and to make this visible in the fields we would need a lot more detailed data.

5.4   LINEAGE 3

5.4.1   Concession HSL-Zuid

NS was the sole railway operator in the Netherlands until about 1996. At this time the first deregulation experiments were initiated. However, this didn't result in NS losing its monopoly position. During the building of HSL-Zuid the Dutch government starts thinking about who could operate the line, and how to generate income to cover the costs of building HSL-Zuid when it is ready as planned in 2005. This lineage covers the events in the competition for the concession to operate HSL-Zuid. The lineage consists of eight fields. The first field concerns the movement from monopoly to tendering the concession; the second and third are about the tendering process; the fourth is about prices agreed upon in the concession; the fifth is about servicing cities as in the concession; the sixth is about renegotiation of the concession; the seventh is again about servicing certain cities; and the eighth field is about the concession with no properly functioning high-speed trains. The different events in the lineage are visualized in Figure 5.10.

Field 1 – from favoured party to one of many (5 February 1998–17 November 1999)
The Minister of Transport announces the idea of dividing the Dutch railway network into different parcels (national and regional networks)

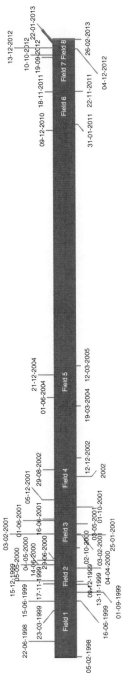

Figure 5.10 Overview of events, lineage 3

during February 1998. This means that HSL-Zuid could be operated by parties other than the monopolist that currently operates the full Dutch railway network, that is, NS. However, and quite understandably, NS is of exactly the opposite opinion and wants the sole right to operate the HSL-Zuid. It considers applying for stock market registration to finance the concession for the HSL. The Minister of Transport is considering this option, too; she adjusts her opinion and is willing to give the exclusive right to operate HSL-Zuid as part of the central national railway network to NS for an unknown sum. The Minister's idea is to register NS on the stock market to generate income. She will put the regional railway networks up for tender if NS does not want to operate them any more. NS may make an exclusive offer for operating HSL, and other parties will only be considered if that offer doesn't meet the requirements of the Minister of Transport. While the Minister is informing the House of Representatives about this procedure several political parties express their doubts, as they are of the opinion that the Minister may be a bit too pro-active and single-minded on this; she should also entertain other options.

Soon after, the Minister asks NS to withdraw its exclusive offer and to participate in a formal tendering procedure. She makes clear that, even though a formal tendering process will be started, the tendering requirements will make sure that NS has a favourable position. One requirement is that of reciprocity; that is, other contenders for the concession have to open up their home railway network for NS, implying that only NS and British parties are eligible, because of the openness of the UK railway market.

Notwithstanding this request, NS makes its official offer for the concession of HSL-Zuid. A committee appointed by the Minister reviews the bid. The bid is declared inadmissible, as it doesn't comply with the requirements of the Minister; that is, NS made a joint bid for the Dutch and foreign part of HSL-Zuid, which should have been two separate bids, and the offer was as least €360 million too low for the whole length of the concession. NS is allowed to make a renewed bid based on certain directives given by the Minister. The House of Representatives is disappointed that the Minister does not inform it sufficiently about the negotiations between the Minister and NS. Maybe owing to the lack of information but also to differing opinions on tendering in general the House is divided about who should get the concession. Some political parties really want NS to win no matter what, others are indifferent, while again some parties do not want a tendering process at all because of doubts about introducing market mechanisms into public services. No matter what the verdict of the House would be, NS suddenly withdraws from the negotiation, as it is not willing to follow the dictate of the Minister. The immediate and retaliating reaction by the

Minister is that she intends to start an official public tendering process for the operation of HSL-Zuid.

Field 2 – disinformation, but still tendering (18 November 1999–4 May 2000)

Since NS has dropped out as the sole bidder, six parties give notice to the Minister that they want to operate HSL-Zuid. These six (foreign) parties are:

- French CGEA;
- the consortium that already runs the Thalys from Amsterdam to Paris, that is, NS, Belgian NMBS, French SNCF and German Deutsche Bahn;
- German Deutsche Bahn together with German airline Lufthansa, possibly with French SNCF;
- British Arriva;
- British Stagecoach;
- British Virgin Trains.

Rumours start spreading that the Minister is debarring NS from getting the concession. The House of Representatives wants an explanation from the Minister about her position towards NS. The Minister informs the House on the content of the offer by NS, which settles the debate. However, NS then gives its own information to the House of Representatives in a closed meeting. It seems that Minister Netelenbos has given the House different information about the conditions for the exclusive bid by NS compared to the conditions that are in the bidding contract to NS, that is, differences in including the international part of the route, and that NS should provide all information about its operational management to new contenders for the existing international routes. A majority of the House of Representatives therefore vote against a tendering procedure. In reaction the Minister agrees to delay the public tender and first research whether NS can return to being a state-owned company so the Minister can have more control, and whether NS is still eligible to be granted the monopoly to operate HSL-Zuid. In the event that these options are not possible she will set the requirements in the public tender such that only NS can win. She is of the opinion that a consortium of NS, Royal Dutch Airlines (KLM) and Schiphol international airport should operate HSL-Zuid. After some persuasion the consortium is formed and makes a bid for HSL-Zuid. The Minister clearly likes the offer and postpones the procedure for the public tender. However, in a completely unexpected move, the Cabinet decide that NS, KLM and Schiphol cannot make an exclusive offer, as they fear that the European Commission won't allow such an offer as it is not in line with EU regulations on railway markets. Hence, once again the concession for HSL-Zuid is put up to tender.

Field 3 – NS wins public tender for HSL-Zuid (5 May 2000–July/October 2001)

Fuel is added to the fire by the House again, as it is still against the plan to have a public tender for HSL-Zuid. It is afraid to lose NS as the provider and that in the tender a large foreign contender might make a better offer than NS and take over the Dutch railways. Shortly after the House has expressed its fear, Deutsche Bahn announces that together with Arriva it will make a bid for the concession. The Minister of Transport partly complies with the wishes and fears of the House. She persists in having a public tender, but will formulate the tender in such a way that NS will most likely be the winner. Two of the most important requirements in the tender are: (1) that every contender must have an office in the Netherlands, because they must have knowledge of the Dutch railway system; and (2) that there should be a level playing field, that is, the contender should come from a country where Dutch transport companies have access to the national railways.

As European legislation has been revised a negotiated contract and exclusive offer by NS are no longer legally allowed. Thus the House of Representatives has to allow the Minister to start the public tender, even though it is still not really confident that NS will win. In total 14 contenders now show their interest in operating HSL-Zuid. These 14 contenders, all consortia, range from foreign train operators such as Deutsche Bahn (DB) together with Arriva, French SNCF, Belgian NMBS and several British, Swedish and French companies, to non-train operators such as the Public Transport Company of Amsterdam. Only four out of these 14 will go through to the second round of the tendering process. DB and Arriva present themselves as serious contenders for the concession. This makes the House of Representatives slightly worried because of reluctance to have foreign operators, in particular DB, on the network. Once the official bidding process has started four candidates make a formal offer before the deadline of 15 September 2001:

- the Dutch/British consortium NS, KLM and National Express;
- the Dutch, French and Swedish consortium Connexxion, CGEA and SJ International;
- the German/British consortium DB and Arriva NL;
- UK railway company Stagecoach.

These offers are evaluated by an independent committee to make sure the most objective decision can be made and to remove any possibility of bias towards certain parties. Based on the list of quality criteria of the tender all four offers are admissible. The four candidates receive the bidding document with all the requirements for operating HSL-Zuid. Meanwhile the Minister makes clear to the House of Representatives, and therefore it is publicly known, that the concession should be at least €100 million per year over a contract period of 15 years. If none of the parties can make a proper offer the state will start its own operator for HSL-Zuid. This provokes a reaction by Deutsche Bahn, in line with its frequent tactics of announcing things in the press, that it finds the requirements for the concession by the Dutch government unreasonable. It suspects that the high cost of €100 million for the concession is simply determined by the high building costs and doesn't bear any relationship to the possibility of recouping the costs through revenues. Meanwhile Stagecoach withdraws as a contender for the concession. No official reason is given, but it is speculated in the press that it couldn't find a proper Dutch partner.

Two weeks earlier than expected the Minister writes a letter to the House of Representatives in which it is stated that the contract for the operation of HSL-Zuid will probably go to NS and KLM. She doesn't explain why the other parties are not in the negotiation process any more, but states that if the negotiations with NS fail she will talk to the next in line, Connexxion, CGEA and SJ, or third-placed DB and Arriva. The offers for the concession made by the consortia are NS offering €178 million, Connexxion €61 million, and DB and Arriva €100 million. On 10 July an agreement on the main issues for the concession of HSL-Zuid is made between the Minister and the High-Speed Alliance (HSA), the consortium of NS and KLM. As the Minister fears that NS has made too high a bid and cannot meet it, she wants to negotiate on the concession; eventually it is concluded that the concession will cost NS €148 million per year for 15 years, providing 96 services per day connecting several cities.

Field 4 – price disputes (5 December 2001–12 December 2002)
As NS needs to pay the high cost of the concession it announces that it will charge higher train ticket prices on HSL-Zuid, about 50–60 per cent more than on the comparable route Amsterdam–Leiden–Rotterdam. The House of Representatives wants the Minister to prevent these price increases and even wants the Minister to write an account of the intended high prices.

NS (HSA) wants to order trains to service the HSL-Zuid. However, before it starts a formal acquisition procedure it wants clarity from the Minister about the options it offered in the bid for the concession. The two options were: (1) €148 million per year and high ticket prices (approximately 50–60 per cent higher compared to the normal train tickets); (2)

€101 million per year, lower ticket prices (approximately 25 per cent higher compared to the normal train tickets) and more seats per train.

Even though the House of Representatives wants the Minister to renegotiate with NS about the price of tickets, the Minister of Transport makes it clear he wants to stick to the first option. NS claims that a lower price for the concession will provide it with the opportunity to apply lower train ticket prices, which will raise the number of train journeys on the route. However, the Minister perseveres in wanting €148 million with no price ceiling on train tickets. Seeing that it has lost the battle the House asks the Minister to request more information from NS to judge if it is really impossible to have a price ceiling for the tickets as is claimed. The Minister promises to look into it.

Field 5 – lost time and lost connections between the Netherlands and Belgium (19 March 2004–12 March 2005)

In spring 2004 it is announced that the travelling time on the Belgian part of the HSL-Zuid will be 17 minutes longer than promised. It turns out that the Belgians have made an error in their calculations. In addition, investments in the track between Antwerp and Brussels have been postponed till 2012. NS immediately reacts by claiming a reduction in its concession cost, as its trains will now have longer travelling times. However, the Minister is not sensitive to this argument and will not reduce the concession fee, as the error is the Belgians'. And on top of it all the Belgian railway company NMBS refuses to run trains to Breda and Den Haag, as it fears that they would be run at a constant loss. However, in the concession agreement it is stated that Breda and Den Haag have to be 'serviced'. This means that NS and NMBS need to find a solution together.

The new Dutch Minister of Transport, Minister Peijs, and the Belgian Minister of Transport, Minister Vande Lanotte, have agreed that NMBS should find a way to give the high-speed trains on HSL-Zuid right of way on the tracks so as to recover some of the lost travelling time (two to six minutes). Later in the year the two ministers are in heavy dispute about the operation of HSL-Zuid, as NMBS still does not want to service Breda and Den Haag. Minister Vande Lanotte should put more pressure on NMBS according to Minister Peijs. The conflict continues, as the Dutch government does not give permission to deepen the Westerschelde estuary in the Netherlands. The Belgians sense retaliation and accuse the Dutch government of doing this because the Belgian government is not applying enough pressure to retrieve lost time on HSL-Zuid and the servicing of Breda and Den Haag. Ultimately the two ministers reach an agreement on HSL-Zuid: Minister Peijs accepts a loss of eight minutes' travelling time in exchange for shuttle services to Breda and Den Haag.

Interim – solutions to price disputes and lost time, but also input for the next field, that is, servicing Breda and Den Haag (23 June 2006–30 January 2009)
During these years several events occur that are outcomes of some of the aforementioned fields, as well as starting input for a later field. However, during these events there were no collective decisions.

The Authority for Consumers and Markets is investigating a complaint that NS is using its monopoly power by taking the Amsterdam–Rotterdam–Brussels train service off its regular service in order to force passengers on to the high-speed train service servicing the same line. Even though Dutch Minister Peijs and Belgian Minister Vande Lanotte had reached an agreement on the services to Breda and Den Haag it turns out that NS and NMBS are not able to reach agreement about those same services. Owing to the delayed realization of the HSL-Zuid (see lineage 2) the Dutch government compensates NS with €3.5 million. Notwithstanding this compensation, NS's financial problems continue to grow as the constant delay in operating HSL-Zuid causes major financial shortages and the Minister is looking for solutions. The outcome is that the Minister agrees with NS that it will receive a 40 per cent reduction to the user fee as long as the Belgian part of HSL-Zuid is not realized. There is also compensation because trains have to go more slowly in Belgium; the cost for the Minister is €16.5 million per year. The Minister considers claiming against the Belgian government for €200 million because Belgium is still refusing to service Breda and Den Haag.

Field 6 – NS uses monopoly power, but still needs to be saved from bankruptcy (9 December 2010–18 November 2011)
Even though NS has reduced prices significantly and has decommissioned the Den Haag Intercity service, travellers still ignore the HSL and use the regular slower but cheaper services. The constant lack of passengers and also the late delivery of the high-speed train V250 (see lineage 4) push NS towards bankruptcy. NS delivers a new business case with a lower fee for the concession to the Minister. The House of Representatives is afraid that this will cost the state several million euros. NS anticipates that the operation of HSL-Zuid will run at a loss for the next 15 years. One of the reasons is the surprising difference in the cost of the Dutch part of the HSL compared to the Belgian and French part: HSL-Zuid costs €130 per kilometre, whereas in France it costs €19 per kilometre and in Belgium €30 per kilometre. These major differences are mainly due to the fact that in the Netherlands it has been a completely commercial tendering process, whereas in the other countries the governments continue to fund the operation through several financial schemes.

To attract more passengers and generate more income NS lowers the surcharge. Owing to the still delayed V250 trains the schedule for HSL-Zuid runs six trains per hour from 2012: four service only the Netherlands, and two are international, that is, one to Antwerp and one to Brussels. The regular train service between Amsterdam and Brussels is taken out of commission as the high-speed rail service takes over that service. Several traveller organizations (Maatschappij Voor Beter OV and Rover) file complaints at the Authority for Consumers and Markets that NS is using its monopoly position to force travellers into the more expensive high-speed rail service. ACM investigates the complaint.

Eventually the Minister has to save NS from bankruptcy, as it is suffering a loss of €386 000 per day on the dysfunctional HSL service. Minister Schultz van Haegen grants NS a renewed concession for the Dutch main railway network. Part of this main railway network is the HSL-Zuid. NS pays €2.2 billion for the concession of the main railway network including HSL-Zuid for 2015–2025. Thus NS will be the main operator of the Dutch main railway network, including HSL-Zuid, and HSA will eventually be merged into NS. The loss for the Minister of Transport is €390 million, which is substantially less than the €2.4 billion in the event of the bankruptcy of NS (HSA).

Field 7 – finally trains going to Breda and Den Haag (22 November 2011– 10 December 2012)
Minister Schultz van Haegen refuses to approve a service schedule for trains on HSL-Zuid as long as NMBS and NS are not performing as agreed, that is, not providing a direct train service between Den Haag and Brussels in accordance with the contract. NMBS should have bought an extra train for this service, but it refuses, because it doesn't think this service is profitable. NMBS also needs to buy extra trains to service Breda and Den Haag. In order to buy these trains, it requests a contribution from the Belgian government, as it is the Belgian government that made the agreement with the Netherlands. Minister Schultz van Haegen is even threatening legal steps against the Belgian government, as it is refusing to invest in the alternative rolling stock for operating HSL-Zuid.

NS announces that starting on 1 January 2013 the surcharge for tickets for the high-speed rail service will be fixed instead of variable and that there will be no extra charge between Amsterdam and Schiphol airport. NS cannot make any adjustments on the international service, as the Dutch and Belgian governments are still in disagreement about the purchase of rolling stock alternatives for HSL-Zuid. However, it is agreed between the Dutch Minister of Transport, Minister Mansveld, and NS and NMBS that a high-speed train will run between Breda and Antwerp. Den Haag will be serviced by an Intercity to Rotterdam, where it will connect with HSL-Zuid. The Dutch government will contribute funds to make this possible, at an estimated cost of between €2.5 million and €3.5 million. NS and NMBS are happy, but for instance traveller organizations Rover (Netherlands) and TreinTramBus (Belgium) are dissatisfied, as it is a compromise between nations which won't benefit citizens.

Field 8 – back to square 1 (13 December 2012–26 February 2013)
Minister Mansveld is clearly not amused by the troubles with the V250 Fyra service, which started running on 9 December 2012 (see lineage 4 outcome for an explanation of Fyra), that is, no direct connections to Den Haag and the required seat reservation. For now, she accounts for these issues as start-up problems.

Meanwhile the Minister admits that mistakes have been made in the negotiations on HSL-Zuid between the Dutch and Belgian governments, as some of the agreements are actually not legally binding. She had an agreement that Den Haag would be connected to the high-speed rail connection between Breda and Brussels, but it turns out that Belgian railway operator NMBS did not sign the agreement. The agreement should have been signed by NMBS to be binding. Thus the claim that the Dutch government is making against the Belgians will probably have no chance of succeeding. However, the Minister investigates all the options for getting a direct connection between Den Haag and Brussels. She will hold NS and NMBS responsible if they do not fulfil the agreements of the concession about punctuality and passenger satisfaction. Failure to comply will result in penalties.

After the failure of the V250 trains (see the lineage 4 outcome) replacement Intercities service the HSL-Zuid. The House of Representatives would like the Intercity train to run more frequently and its schedule to be monitored. The temporary service operating between Den Haag and Brussels will probably continue. The Minister hopes that the V250 Fyra trains will be in operation again; that is still the goal, even though she thinks the chances that the train sets will be fixed are quickly disappearing. According to her, NS and NMBS are working out several scenarios for servicing HSL-Zuid again. She will not consider new contracts, or a new tender for HSL-Zuid.

With these decisions the field, and lineage, ends. Ultimately the train services are the same as in the first offer of NS to service HSL-Zuid, that is, the original plan from 1999, called Intercity Max, which was at that time rejected by Minister of Transport Netelenbos. The irony.

5.4.2 Fitness Fields of the Concession HSL-Zuid

The narrative of this third lineage, that is, all the connected events and fields, looks like the overview in Figure 5.11.

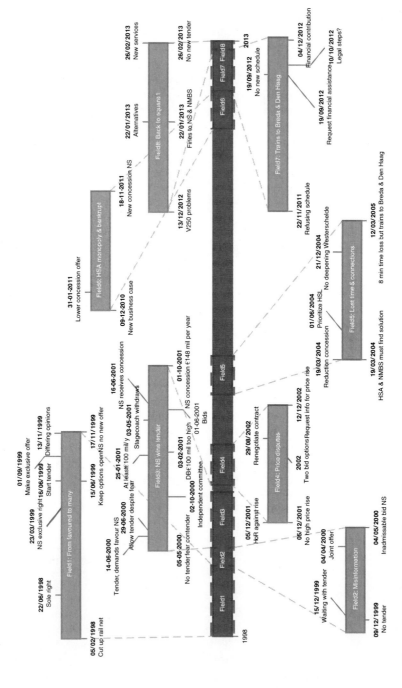

Figure 5.11 Extended lineage 3

The fitness attribution for the respective actors in Figures 5.12 to 5.19 is based on the following argumentation:

In field 1 the House of Representatives (actor 10) is divided in its solutions; it wants the Minister to go more slowly and is partially successful, that is, a fitness of 0.5. The Minister of Transport (actor 11) realizes all her solutions, that is, a fitness value of 1. The NS has major problems realizing its PSD and succeeds only to a limited extent, that is, a fitness of 0.25 (Figure 5.12).

In the second field the House of Representatives is against a public tender, as it has been misinformed by the Minister, but the Minister pushes the tender through, that is, a fitness of 0. The Minister of Transport realizes some of her definitions, and her preferred consortium bids, but fails partly, as the tender will go public, that is, a fitness value of 0.8. The NS joins KLM and makes an offer, but it has to wait, even though its offer is favoured, and it has to redo it in the public tender, that is, a fitness of 0.6 (Figure 5.13).

In field 3 the House of Representatives realizes almost all its goals, that is, the Minister starting the tender and no foreign operator, resulting in a fitness attribution of 0.8. The Minister of Transport completely realizes the solution set with the high bid by the Dutch contender NS, that is, a fitness

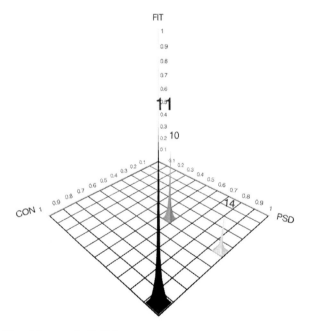

Figure 5.12   Lineage 3, field 1

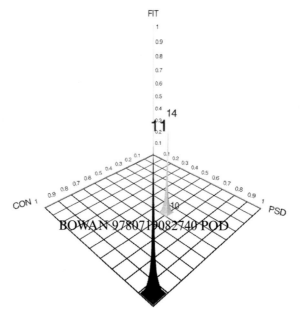

Figure 5.13   Lineage 3, field 2

value of 1. The NS almost completely realizes its PSD; that is, its gets the concession but has to renegotiate, as it knows its bid is too high, that is, a fitness of 0.9. All the other actors (Connexxion, DB and Stagecoach) do not realize their bid being awarded and do not get the concession, that is, a fitness of 0 (Figure 5.14).

In field 4 the House of Representatives wanted a price ceiling, but the decision is that there will be no maximum to the price, that is, a fitness of 0. The Minister of Transport completely realizes the solutions set, that is, a fitness value of 1. The NS offers two options, one of which is rejected, but in the remaining offer it has the opportunity to raise prices as it likes, that is, a fitness of 0.7 (Figure 5.15).

In field 5 the Belgian Minister of Transport once more receives affirmation that the Westerschelde will be deepened, but does not completely get the result with servicing cities, that is, a fitness of 0.8. The Dutch Minister of Transport gets the result that the cities will be serviced again, but with a loss of eight minutes' travelling time, that is, a fitness value of 0.7. NMBS completely realizes its PSD, that is, a fitness of 1. NS does not get its requested lower concession fee, which normally would mean it has a fitness of 0. However, there was a substantial threat of losing its concession, but it was able to maintain it, that is, a fitness of 0.1 (Figure 5.16).

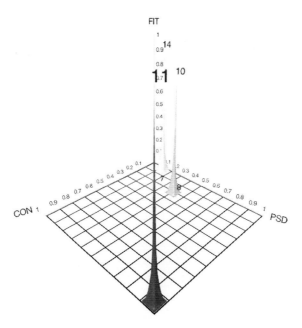

Figure 5.14   Lineage 3, field 3

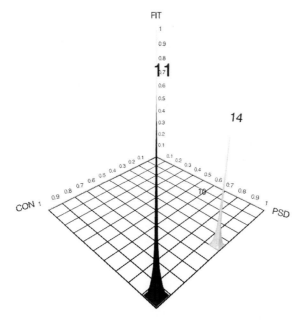

Figure 5.15   Lineage 3, field 4

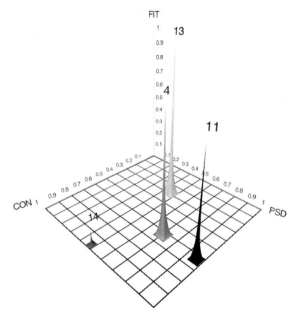

Figure 5.16   Lineage 3, field 5

In the sixth field the Minister of Transport saves NS from bankruptcy, which means the railway network in the Netherlands, including HSL-Zuid, will still be serviced, but pays a price of €390 million for this, that is, a fitness value of 0.6. NS is saved from bankruptcy and remains the prime operator for the railway network, including HSL-Zuid, and has a new concession for a lower fee, that is, a fitness of 1 (Figure 5.17).

In field 7 the Minister of Transport finally gets the requested services to Breda and Den Haag, but has to contribute financially, that is, a fitness value of 0.7. NMBS almost completely realizes its PSD, as it is partially compensated, that is, a fitness of 0.9 (Figure 5.18).

In field 8 the Minister of Transport realizes parts of her PSD, as services to the cities are restored, but there is no high-speed and she has to accept a new train schedule, that is, a fitness value of 0.5. NS realizes its PSD, as it has a new schedule, but it is at a cost, that is, a fitness of 0.6 (Figure 5.19).

5.4.3   Observations Regarding Lineage 3

Lineage 3 is far more dynamic than the first two lineages. Arguably there are more fields in this lineage as more actors mean that more diverging

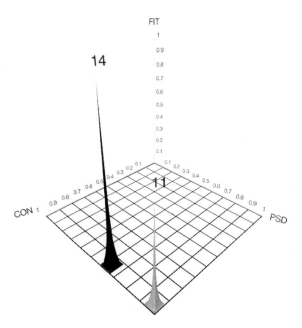

Figure 5.17   Lineage 3, field 6

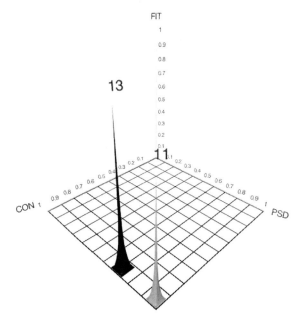

Figure 5.18   Lineage 3, field 7

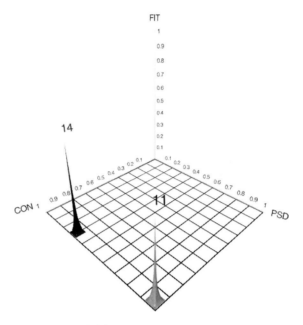

Figure 5.19   Lineage 3, field 8

problem and solution definitions come to the fore. The presence and thus visibility of these actors in the different fields may be due to the fact that this lineage has had much more media coverage and therefore we, as researchers, could get our hands on much more detailed information about the events in this lineage.

Field 3 in this lineage resembles the field in the first lineage, which shouldn't come as a surprise, as they are both tendering processes. This clearly shows in the position of the actors, that is, a clustering of competitors aiming for the same bid, and the tenderer in the right corner. The tenderer has all the connections, while the others are in competition. The difference between this field and the tender field in the first lineage is that the builders in lineage 1 had made joint agreements (recall the fraud), which gave them higher fitness values, whereas in field 3 of the third lineage the train builders are real competitors with one winner and the rest losers.

Two actors are of special interest in this lineage, that is, the Minister of Transport and the NS. First of all, the Minister of Transport is *the* central actor, just as in the previous lineage. Again, this central position makes sense for many reasons. She is the key actor starting the concession; she puts it out to public tender. The Minister can do this as the owner of the tracks, for which the decisions were made in the previous two lineages. As

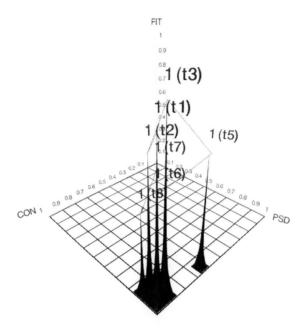

Figure 5.20   Field dynamics of the Minister of Transport

track owner and the one putting it out to tender, she also has a clear vision of what she wants, what kind of requirements are logical, what contenders should offer, and so on. Also owing to the nature of being the owner and seller of the infrastructure, she has to solve every minor or major incident, alone or with others. This means that she is connected to all the actors, but also that she has the most elements from the other actors in her problem and solution definitions. The Minister is able to force or manage, whatever term you prefer, the requirements quite well, and as central actor is able to realize many of her problem and solution definitions; that is, she is able to reach high fitness values throughout the lineage. These dynamics are visualized in Figure 5.20 (which ranges from t1, which is field 1, to t8, which is field 8).

Whereas the Minister occupies a stable position in the corner, NS moves around quite a bit. Sometimes NS is in the lead, for example in the bidding process, but sometimes it is only reacting, as it is for instance when it is saved from bankruptcy by the Minister. This means it is sometimes connected to many actors, while in another field it is connected to fewer actors, and it sometimes has many and sometimes hardly any elements in its PSDs depending on the situation. Its fitness values also range from low to high. Arguably there is progression in realizing more elements in its PSDs for

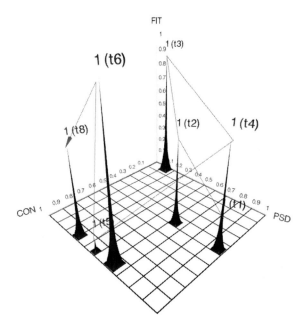

Figure 5.21   Field dynamics of NS

NS, as it gains more success through the lineage, except for t5, where it is bottom (Figure 5.21).

5.5   LINEAGE 4

5.5.1   Operation of HSL-Zuid

While decision making regarding the concession is ongoing (e.g. about the type of service and at what costs), the new concession holder also needs new rolling stock to service the HSL-Zuid and will thus have to find a train builder that can build the required trains. This is the first field of the fourth lineage. The second and last field of the lineage is about problems with building ERTMS into the trains. The different events in the lineage are visualized in Figure 5.22.

Field 1 – tendering rolling stock (19 April 2002–25 May 2004)
NS (HSA) and NMBS need high-speed trains to fully service the HSL-Zuid when the tracks are ready for operation. They start a European tender for electric passenger trains on 19 April 2002. In the tender they ask

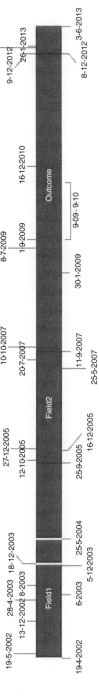

Figure 5.22   Overview of events, lineage 4

121

for 16 trains and an option for 10 extra trains, based on several criteria, of which 'maximum speed of at least 220 km/h', 'price per seat' and 'between 525 and 575 seats per train' are the most important. The trains should be ready for use in October 2006. Seven builders are interested; six of these were already on the shortlist of NS, and there was one relatively unknown builder: AnsaldoBreda from Italy. Eventually four builders submit a proposal:

- French Alstom offers five double-decker coaches (AGV) put into motion by two locomotives (Primas). Alstom cannot offer the TGV owing to a contract with SNCF that operates the Thalys on the same track.
- Canadian Bombardier offers a locomotive that can run at 200 km/h and a variable number of coaches.
- German Siemens offers its Velaro, which is twice the price compared to the 220 versions, because it can run at 300 km/h.
- Italian AnsaldoBreda is the only company offering a newly designed train, with 546 seats, which can run at 220 km/h.

The offer by Siemens is not admissible, as it offers 26 trains in one package, which does not conform to the requirements in the tender. Siemens consciously offered them as one package, because it wants to recover the high fixed costs and development costs for such a small order. Bombardier has a similar problem; it offers 'standard' trains that can go only at 200 km/h because developing a new 220 km/h variant would require high development costs, and the order by NS and NMBS is too small to recover those costs. In short, two candidates remain with trains capable of running at 220 km/h: Alstom and AnsaldoBreda. NS and NMBS prefer the offer by Alstom, as it has the best production schedule and a slightly better price. In addition, even though the AnsaldoBreda trains go at 220 km/h they will not make the projected travelling time between Amsterdam and Brussels. The proposals of both companies are thoroughly checked against 1750 requirements and wishes, both technical and commercial, and the companies are asked to make new offers during the summer of 2003.

The first adjustment made by AnsaldoBreda sees an increase of the maximum speed to 230 km/h, which means that – if it works as planned – the trains will be able to make the required travelling time between Amsterdam and Brussels. It also lowers the price per piece. Alstom offers trains with more seats and a more luxurious package, which raises the price a little bit. NS and NMBS now prefer the offer by AnsaldoBreda over the one by Alstom, a decision that is perhaps influenced by persisting rumours that Alstom is on the verge of bankruptcy.

At the same time the Minister of Transport, Minister Peijs, receives a letter from NS in which it states its intention to order trains that will have a top speed of 220 km/h. This triggers a reaction by her, as she is of the opinion that a travel speed of 220 km/h will not suffice and that 250 km/h is needed to be able to make the contracted travel times between Amsterdam and Brussels. If NS is not able to meet those times the Minister will hold it responsible and penalize it. However, NS and NMBS want to push the 220 km/h trains, as those trains are half the cost of trains that can go at 300 km/h with only a small gain in travel times. They ask AnsaldoBreda if it could build trains that can go faster. AnsaldoBreda complies immediately.

Both companies are asked to make their best and final offer on 5 December 2003. The offer is now for 23 trains, 17 for NS and 6 for NMBS, and 3 trains are still optional. During negotiations close to the deadline, AnsaldoBreda offers an alternative that supposedly can run at a maximum 250 km/h, that is, the now-infamous V250. The new V250 will only cost €300 000 more than the V230; one full V250 train would cost €19 million. For the V250 the price per seat is €34 631, while the price per seat in the AGV by Alstom is €34 384. Both hand in their offers. However, on 10 December, five days after the deadline, AnsaldoBreda makes a small adjustment to the design by removing one toilet and gaining room for 11 extra seats. This lowers the price of the trains by €25 000 and increases the number of seats (thus lowering the price per seat). However, two weeks after the deadline NS and NBMS decide that the expected passenger volumes will require fewer trains and thus order 12 instead of the initial 23 trains. They ask Alstom and AnsaldoBreda to make new offers. Alstom, already dissatisfied by how things went with altered offers from AnsaldoBreda, pulls out of the tender, as it finds 12 trains not enough to design something new for. NS announces that together with NMBS it will order 12 V250s to be produced by AnsaldoBreda and to be ready by April 2006. There is an option for another 14 trains.

Field 2 – troubles with ERTMS (25 September 2005–11 September 2007)
Once the trains have been ordered, several issues arise with ERTMS. This is the same safety system that caused problems during the construction of the HSL-Zuid tracks, but, as this lineage is about the rolling stock, the ERTMS problems are about getting ERTMS working in the train sets.

For many years the Thalys high-speed train has been running between Paris and Amsterdam on regular tracks with standard safety systems. The train, operated by a joint venture between SNCF, DB, NMBS and NS, will also use the HSL-Zuid tracks. For safety measures this train needs to work with the selected ERTMS too. Building ERTMS into the Thalys will take about six months longer than expected, which means that there is a considerable risk that operating HSL-Zuid will also be delayed by half a year. NS accuses the Minister of delivering the ERTMS specifications too late, causing the delay for operating HSL-Zuid. NS could not order new trains, as specifications for ERTMS were missing. Building the systems into the V250 train sets will mean a 10-month delay. Besides ERTMS, NS wants to have another safety system as fall-back, as is the case in Germany and France. Minister Peijs refuses, as she claims that two systems 'tumbling over one another' would make the HSL-Zuid less reliable. She still wants NS to meet the contract, and hopes that by leasing locomotives and coaches HSL-Zuid can still be operated on time. According to the Minister, the mutual accusations about the delay will probably end up in a legal procedure.

NS (HSA) will indeed lease Bombardier Traxx locomotives to operate HSL-Zuid until the V250 is ready. Whereas the V250 would save 30 minutes of travelling time, the Traxx version running at 160 km/h will save just 10 minutes. At the same time NS again pushes its wish to have two safety systems at the same time: the new European ERTMS and the de facto Dutch standard ATB-NG. In fact, ERTMS is creating problems all around Europe for high-speed rail, as it is delivered too late everywhere, and when trains do run there are many malfunctions and subsequent delays. These are mostly due to the European Commission in Brussels constantly making alterations to the ERTMS specifications. The House of Representatives requires from the Minister that she resolves the late delivery of HSL trains because of the safety systems within two weeks. Options are leasing trains or adding an extra safety system to the tracks so the Thalys can switch from its regular route to HSL-Zuid without a hitch. Either option must be paid for by NS according to the House. The problems and constant adjusting of ERTMS also extend to the V250. Building ERTMS in the new trains is equally complicated, as there are conflicting laws and regulations for building and operating new trains between the Netherlands, Belgium and the EU. These delays and adjustments will probably make the train more expensive. An independent bureau publishes a report that shows that the problems of development and implementation of ERTMS have been greatly underestimated and that the respective ministers made errors in choice and design of ERTMS. As a result of this report the Minister of Transport, Minister Eurlings, reports to the House of Representatives that he no longer wants a legal battle with NS (HSA). He apologizes and declares that, from now on, honest communication and information exchange will prevail. He also no longer wants hard deadlines, because they don't seem to be effective.

The Court of Audit presents its report requested by the House of Representatives (see the lineage 1 outcome) on 20 July 2007. It reveals that the delivery of the V250 is delayed again, that the expected passenger volume is much lower in its calculations and that ERTMS is causing many problems and delays. The Minister of Transport pleads guilty, and the Minister discusses compensation measures with NS (HSA). He also admits that it is unclear whether the start of HSL-Zuid in December of that year will be realized. The first reason is the already-mentioned ERTMS problems, but there is also excessive dust formation in the Green Heart Tunnel (see also lineage 2, field 3). Besides these building issues, AnsaldoBreda does not deliver the trains on time, and the leased Traxx locomotives are not yet fitted with ERTMS.

Outcome – end of the line? (10 October 2007–8 December 2012)
As was expected, the delivery of the V250 is perpetually delayed and ERTMS is not functioning correctly in the Traxx locomotives. It will take at least another year before the first trains will be running on HSL-Zuid. The Minister of Transport lives up to his promise and financially assists NS because of the late delivery of ERTMS. He also looks into the option of a temporary – easy to build and implement – alternative safety system in the trains and on the tracks besides ERTMS.

Significantly behind schedule, the prototype of the new V250 high-speed train is introduced to the press on 8 July 2009. The trains are named Fyra, which means 'four' in Swedish. This name is chosen as the HSL-Zuid connects four large cities (none of them located in Sweden . . .). AnsaldoBreda soon starts testing the trains, but it takes some time, which makes the exact delivery date of the trains uncertain. However, NS has a different understanding and expects a train each month starting in the last months of 2010. If AnsaldoBreda is not able to deliver the trains by then, NS will require it to pay a fine.

At the end of the summer in 2009 the first Traxx locomotives start running on HSL-Zuid, shortly followed by the Thalys. However, during the festivities for the press event in which the Thalys runs for the first time, the train stalls at the Dutch–Belgian border as a result of conflicting ERTMS issues – all live on television. NS is still losing income on the HSL because of late deliveries, but also because it generates less income than expected, as the trains are clearly not as popular as hoped for. That is, the occupation rate is only 15 per cent, and one out of five trains is delayed during the first half-year. NS tries to get more passengers into the train by offering tickets at very low prices, but that doesn't make a great deal of difference. The average is boosted during rush hours, in which half of the 498 seats are occupied, but outside rush hours some trains have an occupation rate of only 5 per cent. NS doubles the number of trains per hour to two, hoping that it will attract more passengers.

On the evening of 8 December 2012 the normal train servicing the route Amsterdam–Den Haag–Rotterdam–Brussels leaves Amsterdam Central station for the last time. The V250 starts service the next day.

Outcome – an Italian bullet train in snow is asking for trouble (9 December 2012–3 June 2013)
The first V250 Fyra train leaves Amsterdam Central station on the HSL-Zuid track to Brussels on a Sunday. During the first three days 7 out of 27 trains servicing the route arrive more than 30 minutes late. Some don't even leave, as they break down before departure. There are multiple causes why certain trains are delayed; some will not start, others have software problems and in other cases the train drivers need to get used to driving the V250. After a week of misery with the V250 Fyra, the logistics are improved, ranging from help for passengers on the platform to a trainer or coach for the train drivers. It also becomes possible to buy tickets without a reservation. Lastly the train software is updated to prevent the disappearance of the GSM signal, which is necessary for communication between the ERTMS in the train and the track. Notwithstanding all these efforts the trains keep having problems and many delays; the number of trains taken out of service is five times higher than for regular trains, and 55 per cent of the trains don't arrive on time.

However, these problems pale the moment the snow starts falling at the beginning of 2013. On 17 January three V250 Fyra trains are damaged as a result of the snowy conditions. For safety reasons all the V250 Fyra

Source: Photo courtesy of Sven Schlijper.

Figure 5.23   V250 Fyra trains waiting to be towed back to Italy

trains are taken out of service for two days. At the end of the two days, NS decides to keep the trains out of service, as there is too much damaged sheet metal for the trains to run. The Belgian Department for Rail Safety and Interoperability (DVIS) forbids the V250 Fyra trains on its tracks, as it has found a large metal sheet from a V250 lying on a track. Ice collects underneath the train, and the metal sheets fall off, bounce around and damage the understructure of the train. However, the Thalys is also running on this same route through the same snow and has no problems at all with the snowy conditions. Owing to the halting of the V250 Fyra trains there are hardly any train connections between the Netherlands and Belgium for a number of days.

AnsaldoBreda promises to do its utmost to solve the problems as quickly as possible. It sends a lot of technicians to fix the problems. NMBS is not impressed and is already claiming its money back. NS hopes the problems with the train can be solved, because ordering new trains from a different supplier will cost a lot of time and money. The Minister steps in and requires NS and NMBS provide an alternative for the HSL-Zuid service. She also wants a thorough evaluation of why and how things have gone wrong since the introduction of the V250 Fyra trains. NS and NMBS announce that starting on 26 January regular Intercity trains will be put on the HSL-Zuid as a replacement for the problematic V250. They also make clear that AnsaldoBreda must fix all the problems of the V250 Fyra trains within three months. At this time it seems that NS cannot simply cancel the contract; thus it is putting all its efforts into solving the current problems, as it doesn't see any alternatives in the short or medium term. However, in the next three months the Italians are unable to fix the problems, making the V250 a permanent liability. This provides an option for NS to legally terminate the contract with AnsaldoBreda on 3 June 2013.

As mentioned in lineage 3, the new schedule is now very similar to that of the Intercity Max plan from 1999. Thalys is now the only 'bullet' train on the track servicing the Amsterdam–Rotterdam route several times a day, and once or twice a day it goes to Brussels and Lille. A regular service between Amsterdam, Rotterdam and Breda is provided by Traxx locomotives pulling standard carriages with a speed of 160 km/h. This service is called the Intercity Direct. Passengers still have to pay an extra fee for the HSL service.

5.5.2 Fitness Fields of the Concession HSL-Zuid

The narrative of this fourth lineage, that is, all the connected events and fields, looks like the overview shown in Figure 5.24.

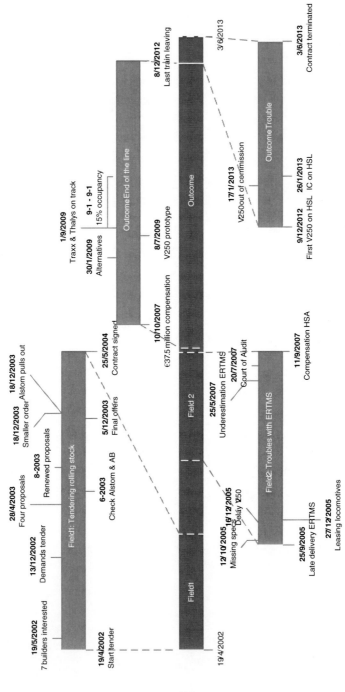

Figure 5.24 Extended lineage 4

The fitness attribution for the respective actors in Figures 5.25 and 5.26 is based on the following argumentation:

In field 1 Alstom remains in the running for the order and realizes parts of its PSD, but in the end loses out to AnsaldoBreda, that is, a fitness of 0.5. Bombardier and Siemens get through to the second round and therefore realize a very small part of their solution definition but in that second round make inadmissible offers, that is, a fitness of 0.1. AnsaldoBreda is the winner of the tender, that is, a fitness of 1. The Minister of Transport is happy that the tender results in her wish, trains that can achieve 250 km/h, that is, a fitness of 1. NS is happy with the contract, but would have liked cheaper trains that could go at 220 km/h. In other words, most of the elements in its PSD are reached, but not all, that is, a fitness of 0.8 (Figure 5.25).

In the second field AnsaldoBreda has big problems building the train, which it can partly attribute to the ERTMS difficulties. That is, it has troubles of its own making, but these are masked to some extent, that is, a fitness of 0.6. The House of Representatives only reaches a small part of its PSD, that is, a fitness of 0.2. The Minister of Transport bounces back and forth between solution definitions and realizes some of them, but also has to compensate NS, as the Minister was at fault in delivery of ERTMS, that is, a fitness of 0.5. NS has major problems delivering the promised service, realizes some of its solutions and gets partly compensated, as it wasn't completely its fault, but in the end it needs to lease locomotives, which costs money, that is, a fitness of 0.4 (Figure 5.26).

In the longer outcome narrative, it is obvious that in due course the winner, AnsaldoBreda, will become a significant loser, as it manufactures trains that are not able to withstand a winter's day, which means that both NS and the Minister are not able to realize any of their PSDs either. This is not taken into consideration for the decision making in the tendering process.

5.5.3 Observations Regarding Lineage 4

Even though there are relatively many actors, especially in the first field, there are not many dynamics. The first field is similar to previous fields in lineages one and two where there is a tendering process; one actor is connected to all others, in this case NS, and has quite a few elements in its PSD as it sets the requirements for the tender. Arguably, AnsaldoBreda has just as many elements in its PSD as it complies with the requirements of NS to show that it is very willing and eager to build the trains, that is, to get the order. Seemingly winners in the first field, AnsaldoBreda has problems staying on top of things, as it is not able to deliver the trains on time, which reduces its fitness. NS also has a significantly lower fitness in

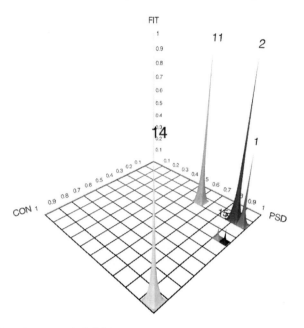

Figure 5.25   Lineage 4, field 1

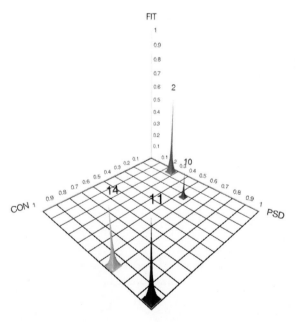

Figure 5.26   Lineage 4, field 2

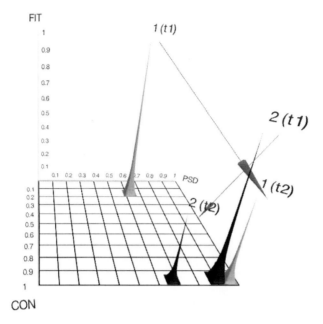

Figure 5.27   Field dynamics of the Minister of Transport (1) and NS (2)

the second field, as it cannot properly service the line, but it still has quite a central position, as it is dealing with many issues simultaneously. Notice the less prominent position of the Minister of Transport in the first field and the prominent position in the second. The Minister is clearly more at a distance, leaving the public tendering for rolling stock to NS and only reacting once to one of the requirements in the tender (i.e. the issue with the maximum speed). This changes drastically as the table turns to problems with ERTMS, for which the Minister is partly responsible. Thus the Minister has to react to many incidents and find solutions to others, hence the move to the central position once more, holding most elements in the PSD. The movement of these two actors is visualized in Figure 5.27.

5.6 COUPLINGS

As was already hinted at in the narratives, there are certain couplings between the different lineages, as there were either several events that impacted configurations simultaneously or one event that impacted other configurations at a later moment in time. There are two asynchronous and three synchronous couplings between several lineages (see Figure 5.28).

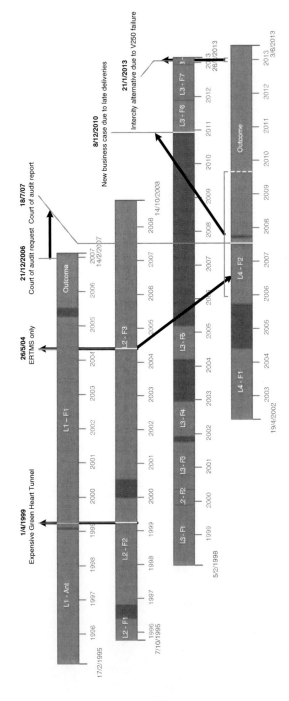

Figure 5.28 Coupled lineages

The different couplings will be discussed in the sequence in which the coupling events occurred in time.

The first synchronous coupling occurs in the second lineage with different scenarios about how to go through the Green Heart. This sets the space of possibilities in the tunnel discussion as well as in the financial discussion. After a number of debates the problem and solution definitions align towards a tunnel. The widely supported tunnel is the longer, drilled version, which is the more expensive option. The high price of the tunnel is defining for the solution definitions by the actors involved in the decision making on the preferred tunnel, but it also has a major impact on the amount of money needed for the completion of the construction of HSL-Zuid. In other words, the discussions about how to finance the building of HSL-Zuid in the first lineage, based on the PSDs of the involved actors, are immediately and significantly impacted by the decisions about the track alignment in the second lineage, and vice versa. The coupling between the lineages creates a mutually positive (reinforcing) effect. However, when keeping the outcome in mind for the government, builders, operators and citizens, the result is mixed, if only because the preservation of the Green Heart is very costly.

The second synchronous coupling occurs when the Minister opts for ERTMS as the sole safety system (lineage 2, field 3). This sets the boundaries for certain solution definitions not only in this lineage, but also in two other lineages. It has an impact on the finances needed for constructing all elements of HSL-Zuid. The Minister needs longer negotiations with for instance Siemens about the costs; thus delivery of ERTMS is delayed. This feeds back into the construction and safety issues of lineage 2. It also feeds forward into lineage 4, field 2, where the failing ERTMS hinders the operation of the Traxx locomotives, Thalys and V250 Fyra. In other words, the event of ERTMS being chosen as sole safety system establishes a coevolutionary relation between lineages 2 and 1, and an asynchronous coupling between lineages 2 and 4. The coevolutionary relationship is of a reinforcing kind, where the delays caused lead to rising costs, and more discussion about the cost means extra delays. Thus it features a similar argumentation to that of the first coupling.

The third coupling is a relatively simple asynchronous coupling. The House of Representatives is concerned about the financial setbacks in the building process of HSL-Zuid and requests independent research (first lineage). This lineage is already concluded at the time the Court of Audit publishes its result, because there is no real collective decision making on finances any more. However, the publication of the report about six months later has significant impact on the fourth lineage, because it makes clear that the expected passenger volume has been greatly overestimated and that the difficulty of implementation of ERTMS has been greatly underestimated. This sets or, better, *re*sets discussions about both ERTMS and how NS is going to service the line without trains readily available. In other words, the request in the first lineage sets the boundaries in the second field of the fourth lineage.

The fourth coupling is between the second field of lineage 4 and the sixth field of lineage 3. The coupling is between the problems with attracting passengers and the delivery of the required high-speed train sets in the fourth lineage, and the collective decision making about the concession in the third lineage. The delivery of the V250 Fyra trains is constantly postponed, and the lack of passengers on the slower alternative trains nearly causes NS to go bankrupt. The long period of being in the red in operating the HSL forces NS to hand in a new business case with a lower fee for the concession on 9 December 2010. Here, the problems in lineage 4 feed forward into the PSDs of the actors in lineage 3.

The final coupling we would like to mention here is similar to the previous one. After the V250 Fyra starts servicing the HSL-Zuid, that is, the outcome of lineage 4, the concession is put under tremendous pressure, as NS is unable to put a proper service on the route, that is, the last field of lineage 3. The PSDs of many of the actors are influenced tremendously by the fact that the trains are not able to service the line. This ultimately leads to the abolishment of the envisioned high-speed rail service and the alternative Intercities being put into service. These trains are only able to perform at a lower speed, but they replace the V250 permanently as the contract between AnsaldoBreda and NS is terminated.

5.7   CONCLUDING REMARKS

The main challenge in this chapter was to show the full details of our model. We have analysed a long journey of connected events of decision making. The string of events was cut into four distinct lineages. The number of fields differed significantly per lineage. Notwithstanding this difference we were able to show how the model helped in understanding the motivations and movements of actors, related to their problem and solution definitions or how they align with others. For instance, the centrality of actors and their societal involvement showed by their remaining in the corner with many elements in their PSDs and a high connection score. In contrast the volatility of NS was visible, as it jumped around the grid in search of success or, better, in order to avoid failure. The evolutionary nature of decision making was substantiated by the dynamics the model was able to unravel. Certain aspects of the dynamics, like the coupling between lineages, make clear how events or decisions set the course in other lineages. Even though this HSL-Zuid study shows many events divided into many lineages and fields, it is in itself a relatively simple study and as such has limited dynamics. To demonstrate the three different versions of dynamics more fully we will focus on them separately in the next chapter.

6. Enter in time: analysing dynamics in three empirical cases

6.1 FOCUSING ON DYNAMICS

The previous chapter showed how structural information about actors, preferences and connections and a detailed narrative can combine to provide a thorough understanding of the intricate complexity of collective decision making. We have seen how coupledness can create unforeseen situations, how lineages can become trapped through reinforcing feedback loops, and how the formal separation between collective decisions and implementation becomes blurred as implementation feeds back or forward into the decision-making process. Along the way, actors have become entangled in a complex web of lineages from which there are no easy escapes.

We can't overstate the importance of time in the emergence of this complexity. Time adds an important dynamical aspect to collective decision making, because it means that actors will meet again (i.e. there is no such thing as a single-shot game), upping the ante for all the people involved. The purpose of this chapter is to zoom in on three main dynamical aspects of collective decision making as highlighted by our model: field-bound dynamics, lineage-bound dynamics, and coupledness between lineages. In order to strengthen the robustness of our findings, we will analyse different data sets from those in the previous chapter.

6.2 FIELD-BOUND DYNAMICS

The analysis of collective decision making sometimes seems to suffer from the so-called Droste effect: a recursive phenomenon where traces of the whole are mirrored in the parts. The model presented in the previous chapter was able to capture the dynamics of time in the shifts *across* fields. However, actors also undertake activities *within one field* in order to seek those PSDs that seem to lead towards improved fitness. When a promising configuration has been found, the actors make their moves accordingly. Other actors may respond to that search process, because they are also in search of those configurations that offer a road towards fitness. These search processes do not always lead directly to the conclusion of a field, yet it is useful to consider them, because much of actors' lives is exactly about the attempts at the micro-level to identify what elements of PSDs are most likely to offer higher fitness. We will tease out these dynamics in a study about the collective decision-making process regarding the future of the Gotthard region in Switzerland.

6.2.1   Mountains of Nothingness? The Gotthard Case

The Alpine Gotthard region in Switzerland is best known for its mountain pass. It connects the north of the country and its southernmost regions or, if you are a tireless traveller, north-west Europe with Italy. The use of the pass can be traced back to medieval times, but the Gotthard rail tunnel constructed in 1882 has been a very important factor. The tunnel's history is a heroic tale of record-breaking building techniques, lost lives and gained prosperity. Indeed, the local communities in the Gotthard region now consider the tunnel as fundamental to their identity (Schueler, 2008). The Gotthard rail tunnel was followed by the Gotthard road tunnel in 1980 for cars, and the massive, 57-kilometre-long Gotthard Base Tunnel (GBT), which opened in June 2016. The latter is set to replace the old and ageing railway tunnel and to facilitate the modal shift of freight from road to rail. Importantly, the new tunnel will allow trains to run directly between Erstfeld and Pollegio without having to climb the steep ramps to Göschenen or Airolo first (Figure 6.1). Subsequently, the travel time across (or rather under) the Gotthard region will be reduced by 45 minutes.

Just in case you were wondering: no, this is not going to be another railway-focused study. We are much more interested in the effects of the impending opening of the GBT on the Gotthard region itself. It is sparsely populated and offers little in terms of economic activities. Views of small villages in fresh Alpine meadows against a backdrop of snowy mountain tops are gorgeous, but it is hard to live in a postcard. The mountains also mean that economically vital places are much further away than a first glance at the map may suggest. The old railway provided a lifeline for the region, connecting its villages as it swayed up the ramps towards the tunnel entrances. Indeed, anyone wanting to travel between the northern and southern part of Switzerland had to use this route. The new tunnel will gobble up most of the railway traffic and additional road-bound freight traffic, and will leave the whole region without much in terms of direct connections to Switzerland's main cities. Additionally, the region fears that the use of the new tunnel will render the region invisible – quite literally so, because most people will now travel underneath it.

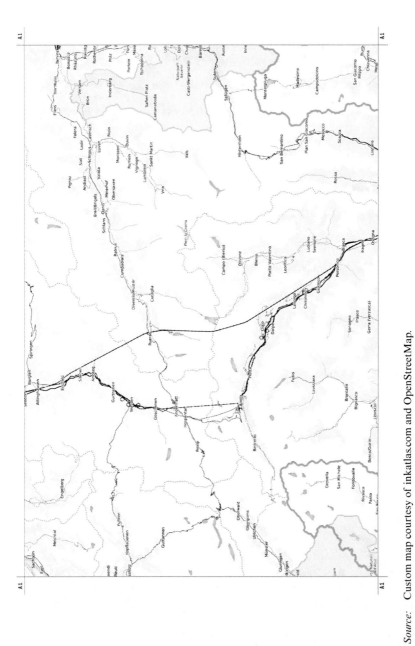

Source: Custom map courtesy of inkatlas.com and OpenStreetMap.

Figure 6.1   Overview of the Gotthard region in Switzerland. The new Gotthard tunnel runs from Erstfeld in the north to Biasca in the south. The location of Porta Alpina is approximately at the Rueras location. The old Gotthard route is located further to the west, with the tracks spiralling up to the old tunnel between Göschenen and Airolo

The imminent opening of the new tunnel led to much soul-searching among the cantons and settlements in the region (see Schueler, 2008 for an extensive antecedent narrative). The start of this process can be traced back to an initiative by national councillor Brigitta Gadient in October 2000. She proposed constructing a station halfway along the GBT to serve the region. The GBT required emergency shafts to the surface, and Gadient suggested that those emergency shafts could double as elevators to move passengers between the Sedrun area and the station 800 metres lower. The project, named Porta Alpina, would require careful timing with the ongoing GBT project and, importantly, major capital investments. The total construction costs were estimated at 50 million Swiss francs as long as the construction of the station was combined with the construction of the GBT. That is pretty steep for a minor station, even by Swiss standards.

Source: Photo courtesy of Georg Trüb.

Figure 6.2   *The Sedrun subterranean station in the new GBT underneath the massive shafts extending through 800 metres of rock. There was a rare opportunity to visit the station during the opening festivities of the tunnel in August 2016. The building of the station never moved beyond the shell construction and will not be made operational*

Perhaps an idea that 'sounds too utopian to be true' (Schueler, 2008: 152), it actually gained considerable momentum among regional actors, leading to requests to the national Bundesrat to approve the project. In response, the Bundesrat agreed at the end of June 2005 that the region was in a vulnerable position but also refused to support and finance Porta Alpina straight away as long as it was not embedded in an overarching strategy for the future of the entire region.

This response proved to be a watershed in the process. The demanding actors understood that Porta Alpina wouldn't solve all the fundamental issues in the region, such as low economic growth and employability, as long as they didn't establish closer cooperation and coordination. Indeed, the first drafts of the strategy required by the Bundesrat stated that:

> [O]wing to the lack of a comprehensive spatial concept for Gotthard, the alternatives are currently just uncoordinated investments in a multitude of activities and projects with relatively little added value. The coordination of activities in the region and the alignment towards further development of a brand 'Gotthard' offer the chance to show a new regional policy in the eco- nomically underdeveloped Alpine region as an alternative. (Bau-, Verkehrs- und Forstdepartement Graubünden, 2005: 5, translation by authors)

The document, drafted by the four cantons and other regional partners, further stated that the 'central task in this *process* until the competition of the Porta Alpina is the establishing of a cooperation of the actors' (ibid., italics in original).

Ultimately, Porta Alpina as a solution faded away slowly because of many practical considerations. Conversely, the idea of a regional strat- egy supported by all the actors gained much momentum as a promis- ing solution. It solidified in a cooperation process, Projekt Raum- und Regionalentwicklung Gotthard (PREGO, Gotthard Spatial and Regional Development Project), and a report, 'San Gottardo: Das Herz der Alpen im Zentrum Europas' (Gotthard: the Alpine heart in the middle of Europe) (PREGO, 2006), released on 31 January 2007. Almost immediately, 'San Gottardo' started to become synonymous with the comprehensive devel- opment plan for the region. The plan, with a time horizon of 2020, covers many regional aspects such as culture, tourism, nature conservation, the job market, the role of municipalities, and public transport. Importantly, the region is treated as one coherent whole instead of as a collection of 'islands'. This is a change from the past, when the common approach was to divide it up by means of administrative boundaries. According to the plan, the region has itself to blame for its economic woes, because 'too little cooperation prevented the creation of mutually beneficial future per- spectives and solution strategies' (PREGO, 2006: 5). The report then adds to the sense of urgency by emphasizing that, without further cooperation and integration, the weakest of the municipalities within the region will not survive. A series of conferences confirmed the ambitions. The plans show a far-reaching attempt to present the region as one, even though it is recognized that its various parts have their own identity. It is a very heterogeneous region, and there are no clear geographical borders, so the cooperation and coordination are driven by the pressing current issues, not because the actors all share the same (cultural or administrative) background. Yet there is a drive towards consensus, as can be expected from the Swiss decision-making culture, which is very democratic and consensus-oriented.

6.2.2   Search Processes within One Field

The Gotthard study is interesting because of the many changes in the respective PSDs as actors search for the definitions and combinations of problems and solutions within the space of possibilities that promise a better fit. The search process starts because none of the actors have a clear understanding of what the winning combination could be. Looking back, it is obvious that the starting point, Porta Alpina, was actually a solution in search of a problem. Over time, it was realized that, first, the problems of the region couldn't be solved by building a very expensive train station and, second, many of the issues were down to the fact that many of the actors didn't cooperate or think alike even though they were subject to the same pressures. Even the meaning of Porta Alpina itself changed over time: from a solution for the problem definition of a vulnerable region to a device through which more cooperation could be fostered. The idea was (slowly) abandoned, as it had become superfluous in the light of further cooperation and regional integration around the San Gottardo manifest.

The study demonstrates the within-field dynamics that can occur when PSDs mutate over time without any clear and immediate gains in fitness for one (or a group) of the actors involved. While strong demarcations in the decision-making process are absent, there is still movement. This is partly in response to external pressure, partly a self-generated cascade effect. The search process and subsequent evolution of PSDs are shown in Table 6.1 and Figure 6.3.

Table 6.1 shows the various manifest problem and solution definitions per event, with time running from left to right within the single field, and the extent of support for each definition in terms of the share of actors sharing a certain definition. The table was generated using the persistence export function in un-code.org. We can see that the number of problem and solution definitions expands over time and that the number of fully shared problem and solution definitions increases accordingly. Mutations

Table 6.1   Persistence and change of PSDs within one field, as distributed across time

Problem Definitions	First initiative Porta Alpina (t1)	Initiative council of Graubunden (t2)	Council of States requests Bundesrat (t3)	Bundesrat requests concept Gotthard (t4)	Release Gotthard development concept (t5)	MoU between cantons signed (t6)	No further pursuing Porta Alpina (t7)	Release PREGO concept San Gottardo (t8)
Development of Tre Valli region						20%		20%
Further postponement of Porta Alpina leads to much higher costs							100%	
Lack of clear strategic concept for the region				100%	100%			
Lack of cooperation among actors in the region					100%			100%
Projected costs of Porta Alpina			100%					
Time pressure pairing Porta Alpina with GBT							100%	40%
Unknown cost/benefit ratio Porta Alpina			100%	100%				
Weak development of tourism						20%		40%
Weak economic situation in Surselva	100%	100%	100%	100%				40%
Weak infrastructure links of region to rest of Switzerland	100%	100%	100%	100%				20%
Weak social-economic structure of the region					100%	100%		100%

Solution Definitions							
Cancel Porta Alpina		100%					
Diversification of the regional economy	100%		100%				
Initiate regional projects such as Parc Adula			20%			100%	
Investigation of possible solutions		100%					
Maintain frequent railway connections with the rest of Switzerland			20%				
More cooperation and coordination among actors in the region	100%	100%	100%	100%			
Promote the region	100%	100%					
Promote tourism	100%	100%	20%				
Share the burden of preparation / shell construction costs				100%			100%
To build a station inside the GBT for the region and to use the emergency exits as access to the station			60%			50%	
To develop a joint strategy for the future of the Gotthard region			80%	100%	100%		
To integrate Porta Alpina into a broader regional concept			100%				
Upgrade connecting infrastructures (e.g. MGB)			40%				

Total no.

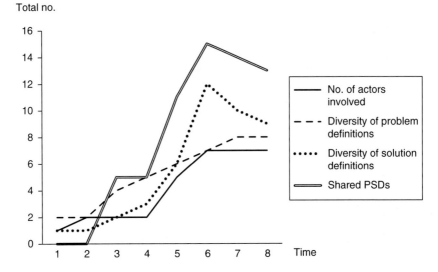

Notes:
The number of actors is cumulative.
Actors that were mentioned in the process but did not have a share in the decision-making process, such as the interest group Visiun Porta Alpina, have been left out.

Figure 6.3   The number of actors within one field in relationship to the diversity of the problem and solution definitions, as well as the number of shared definitions

occur because of a number of events. The first such event is when the Bundesrat responds to the request to fund Porta Alpina by saying that something must happen in the region but building Porta Alpina without a broader development concept for the region is not going to help (t_4–t_5). The other actors respond to this by starting a development process that requires other actors from the region, most prominently the cantons of Ticino, Uri and Valais, to join and cooperate. This is entirely rational from the evolutionary perspective: the actors involved now know that a more comprehensive development plan has a higher probability of attracting funds and people to the region than Porta Alpina as a stand-alone project.

The entry of new actors (t_4–t_5) leads to considerable variation of solution and problem definitions. As shown in Figure 6.3, the higher variety between t_5 and t_6 corresponds directly to a simultaneous increase of *shared* elements in the PSDs. In other words, there is agreement about the additional problem and solution definitions created. Obviously, it is tied to the particular interests of each actor, but it is also a tell-tale sign of the substantive variation that is taking place. None of these interests are killed or ignored but are added to the pool, and most of them are supported by many, if not all, of the actors involved.

The variation is such that Porta Alpina starts disappearing, slowly but surely, from the space of possibilities. Again, its disappearance doesn't happen overnight. It simply becomes less and less important in the overall development concept, which includes many other measures. The ongoing construction of the GBT provides an opportunity to say goodbye to Porta Alpina (t_7), because the window of opportunity to build it at relatively low costs passes before design and finance have been realized. By that time, the actors have already understood that the project wasn't going to salvage their future. We can therefore see considerable consensus about both the cancellation of Porta Alpina and the future direction of the regional cooperation.

The disappearance of Porta Alpina and the PSDs related to that particular project means that the pool of PSDs as well as the number of shared PSDs has decreased by the time PREGO is released (t_8). While presenting a development plan for the whole region, the actors also carefully avoid the suggestion that the region is a coherent whole. Each part of the region has its own particular characteristics and issues to deal with and pursues its own goals in **PREGO**.

The study demonstrates the search processes taking place within *one* field and demonstrates that variation, selection and retention within the space of possibilities are real. The actors probe in multiple directions in that space in order to get an understanding of which PSDs may yield a better result. This exploration corresponds to a substantive expansion. Having generated an image of what is seemingly the best possible PSD, there is no longer a need to maintain an expanded pool of PSDs. A narrower focus on the selected PSDs emerges as actors converge around the set that they believe will improve their fitness. Each actor now gears up for the decision making about the funding and implementation of (parts of) the development plan. That will mark the solution of the field. For now, it is important to understand that each field in a given lineage can be dynamic in itself, as the search process requires actors to exercise variation, selection and retention within the pool of PSDs derived from the space of possibilities.

6.3   LINEAGE-BOUND DYNAMICS

Now that we have teased out the finer details of the dynamics within one field it is time to step up one level and look at the dynamics between fields *within one* lineage. We have already seen in Chapter 5 that collective decision making is not a matter of single-shot games. It is very likely that actors are present across fields – save for those actors that decide, or are forced, to fully leave the arena. This means that gain or loss of fitness in one field can be redeemed in the next field if an actor plays its part well. Such gains or losses can be attributed to the strategies of the actors but also to (random) chance. In fact, a good understanding of chance and the ability to seize the opportunity could contribute to fitness gains. We will now focus on a number of lineage-bound dynamics: how actors reciprocally respond to each other in search of fitness across fields in one lineage. We will demonstrate these cross-field movements using a study of the attempts to build a sports-centred urban district in Rotterdam, the Netherlands.

6.3.1   Sports in the City: The Rotterdam Case

Sports have always been one of the key policy objectives of the municipal authorities of Rotterdam. It has hosted an international and popular marathon each year since 1981, a four-day equestrian event since 1937, a world tennis tournament since 1974, a biennial international baseball tournament since 1984, the world beach volleyball championship, and many more. These sport events are held in different locations in the city. One of those locations is in and around the stadium of Feyenoord professional football club. The stadium is located in the somewhat eccentric IJsselmonde district bordered by a large railway yard in the west, an industrial park nicknamed 'Little Belgium' (because of its ugliness) in the north, a small residential area in the east named after various Olympic disciplines, and a confused heap of social housing, small parks and abandoned hospital in the south.

The area has been subject to redevelopment since the 1990s, albeit in a haphazard way. The football stadium, nicknamed De Kuip (The Bathtub), has a capacity of just over 51 000 seats and as such used to be appointed as the stadium for final matches of big international tournaments. It has also hosted many concerts, for example U2, Metallica and Pink Floyd. The stadium and the sports centre connected to it are widely regarded as important for the region. However, the stadium was built in 1937 and is not up to modern standards any more; for example, it lacks adequate sanitary facilities, lacks capacity for more visitors and sponsors, has inadequate logistics, and so on. Over time, the stadium moved from being one of Europe's top venues to an inferior position, with international games and concerts taking place elsewhere (Lievaart, 2014).

The 1990 renovation of the stadium was a stop-gap measure meant to last for about 20 years. This was the start of an extensive round of discussions between the so-called 'Feyenoord family', a term used for a group consisting of the directors and owners of the stadium and the professional and amateur sections of the football club, and the municipal authorities. However, nothing very substantial happened until the municipal elections of 2006. A new governing coalition was put together, consisting of PvdA (the social democratic party), CDA (the Christian democratic party), VVD (the liberals) and GL (the green party).

This new coalition government aimed to revitalize the whole southern area of Rotterdam – traditionally the poorest district of the city – and that includes the IJsselmonde area. The focal point would be both top and recreational sports, and the stadium area was envisaged as playing an important role in this redevelopment. This resolve coincided with the plan of the Royal Netherlands Football Association (KNVB) and the Royal Belgian Football Association (KBVB) to make a bid for the football World Cup 2018 in the Netherlands and Belgium. Conveniently, a new stadium would land Rotterdam as the host for the opening and final match. The actual lineage for the redevelopment of this area starts here and ends in mid-March 2015 when the decision is made to cancel the plans for a new stadium. The lineage of relevant events, covering six distinct fields, is shown in Figure 6.4.

Even though the idea to seize the momentum was broadly supported, there was a big elephant in the room: there was no budget at all, and the business cases were weak, with one party, the Feyenoord professional football club, on the verge of bankruptcy. The municipal executive was betting on a successful bid for the football World Cup. An accepted bid would probably mean the allocation of considerable funds from the national government. In other words, the whole plan was built on some rather significant assumptions. As we know now, Russia was selected as the host for the 2018 World Cup. This coincided with an increasingly worrying financial status for the city of Rotterdam. The financial and economic crisis starting in 2008 had forced the city's budget deep into the red, and massive budget cuts and lay-offs were made. Under such conditions, the original prestigious district-wide redevelopment was out of the question.

Alternative options had to be pursued, and the Feyenoord family, believing that a new stadium would be a necessary condition for getting out of their financial troubles, tried to forge ahead with the development of a cheaper plan. The municipal executive, while supporting those attempts, also separated the broader development of the district from the building of a new stadium because of the highly controversial nature of the latter. The broad development continued under the header of Stadium Park (the development of the new stadium changed headers often, ranging from 'Super Kuip' to 'Save De Kuip'). The development of the new plans took place with considerable turmoil: coalition parties leaking information, a

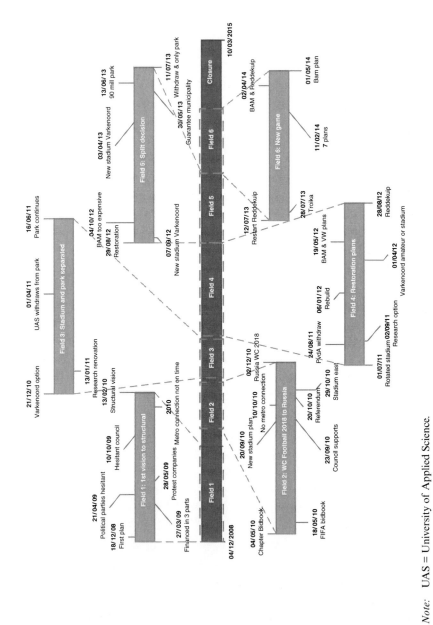

Note: UAS = University of Applied Science.

Figure 6.4 Full lineage of the 'Sports in the City' study

148

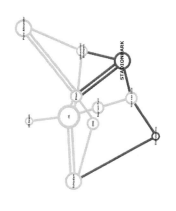

Figure 6.5   Plans of the new metro connection (bottom left) and the new stadium at the Maas, with the old stadium in front and the whole sports campus

municipal election and subsequent change of coalition, and Feyenoord fans protesting against a new stadium because the old one – holy ground in their eyes – would have to be abandoned. Indeed, at some point, some of the fans even managed to develop an alternative plan for the renovation of the old stadium and to get the city council to consider that proposal very seriously. While sections of the Stadium Park crept towards realization, the new stadium didn't. A final push was made by means of a very public tender, but the winning consortium ran into financial troubles itself and needed to crank up the building price of the offer in order to stay afloat. This was outside the budget range, and the plan was cancelled.

6.3.2 Actor Movements across the Fields

The study offers a microscopic view of the ways in which actors move across fields in order to obtain better fitness. There are many stakes here. Political parties look for popular votes, schools look for ways to get their pupils to become more physically active, the Feyenoord family dream of a shiny new stadium, and the fans being fans are passionately against or in favour of everything. They also have voting rights in Rotterdam, so that connects them to local politics, which closes the circle nicely. The overall picture across six fields shows a very dynamic dance across the board (Figure 6.6), where the actors are numbered as follows:

1	BAM	8	PvdA
2	CDA	9	Reddekuip
3	D66	10	SP
4	Feyenoord family	11	VolkerWessels
5	GL	12	VVD
6	LR	13	ZwartsJansma
7	Municipal executive		

The overview in Figure 6.6 shows at a glance that certain actors moved considerably over time, while others stayed put. For example, consider the moves of LR. LR started as an anti-establishment party and did very well in a number of elections. It was an opposition party at the beginning, and it was only natural for it to be against any plan by the municipal executive. Its contrary stance was reinforced because it appealed to a certain segment of its voters: Feyenoord fans who were vehemently against any plan that would repurpose the old stadium. The 2014 elections, however, landed it a seat in the governing coalition, and that compelled it to adopt the existing stance of the municipal executive. This was not just a matter of sticking to the habits of Dutch local government but also because the solution

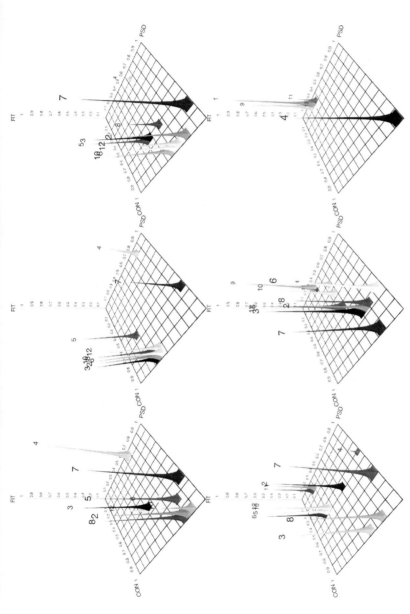

Note: The final field shows a limited number of actors because it concerns the discussion between the Feyenoord family and two competing offers for the stadium.

Figure 6.6   Movement of the main actors across the six fields in the 'Sports in the City' study

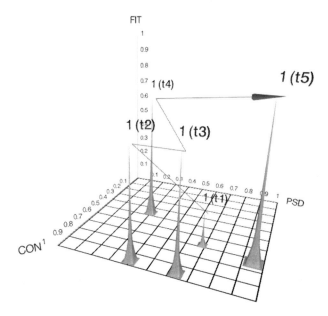

Figure 6.7   Actor LR moving back and forth across the landscape because
of the changes in the nature of PSDs present at given time
frames

space had altered over time – from a comprehensive development plan
largely paid for by tax money to a proposal where the Feyenoord family
had to present a business case on the basis of private money. The change
from opposition to co-optation is shown in Figure 6.7. Such a movement
shows that becoming a member of a coalition does have real effects on the
position of an actor in relationship to others.

   Naturally, our model highlights the mutual dependence between actors
and the fact that the position of one actor is relative to the positions of
others. If one moves, the others respond by moving along or by attempt-
ing to keep position. Such mutual adjustments can be found between the
municipal executive on the one hand and the Feyenoord family on the
other. Their relationship is complicated. Both actors need each other, as
the development of Stadium Park and the new stadium are deeply intertwined. But they are also a little weary of each other: the Feyenoord family
expect the municipal executive to help with financing, and the executive
believes that the Feyenoord family should manage their own business.
But then again, if left to their own devices, a new stadium is going to be
unlikely, because the Feyenoord family haven't got enough clout to make
it happen. A failure to get a new stadium is also not in the interest of the

municipal executive. As time moved on, the two actors were increasingly portrayed in the media as being diametrically opposed. In reality, however, their interests were too intertwined for the relationship to be cut. In their joint search for fitness, they move in tandem. Their mutual movements are shown in Figure 6.8.

If we take a closer look at the tandem movement of the Feyenoord family and the municipal executive, we notice that there are some subtle but important differences between them. A particularly important point is the considerable difference in fitness values (see Figure 6.9).

Figure 6.9 shows that the Feyenoord family were close to realizing their goal, which was to build a new stadium, to be ready in time for the 2018 World Cup. Following the financial crisis and the loss of the bid, it can't continue in the same way as before. It took a while for the Feyenoord family to realize this and to slowly regain fitness (after t_4). Being so dependent on the municipal executive, they *follow* the preferences and course set out by the executive (which also needs to adapt to changing circumstances, of course). In other words, the municipal executive acts as a conduit for the Feyenoord family to understand what elements in the PSD and strategy are needed to gain fitness. Thus we can see how the adaptive movements of both actors converge over time and that the Feyenoord family regain fitness as a result (see also Figure 6.8). A caveat is in order here: the selected plan turned out to be too expensive and was rejected after the decision had been made to build it. In other words, the Feyenoord family and the municipal executive (to a lesser extent) were king of the hill for only a day. However, this transpired after we had concluded our analysis.

Let's turn to another aspect: the complicated relationships between actors. Network analysis often has a challenge in differentiating between various kinds of connections: in most versions, actors are simply considered to be either connected or not. One can describe the nature of the connections but often in a rather rudimentary fashion. Political parties, for example, are obviously all connected, because they are mutually dependent in the collective decision-making process. At the same time, however, they do not always agree. But that disagreement is not just a simple matter of a difference between government and opposition. The model can show the complex relationships between actors by using the PSDs as a differentiator. The fifth field of this lineage serves as an example. This is when the municipal executive and the Feyenoord family make a last big push in order to get enough votes for their plans (see Figure 6.6).

As can be seen from Figure 6.10, the introduction of the substantive dimension helps differentiate the constellation of the actors on both their connections and their (dis)agreement over the subject of the decision-making process. For example, we notice that GL (3), SP (6) and VVD (7)

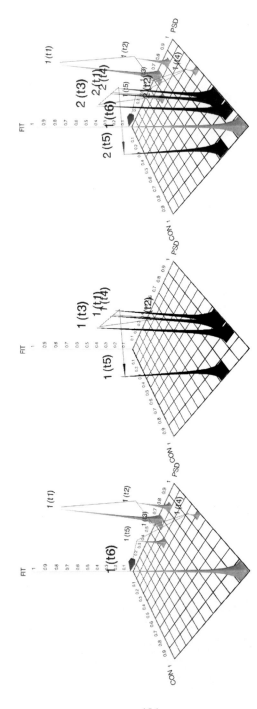

Figure 6.8   The Feyenoord family (left) and the municipal executive (middle) moved in tandem across the fields (right)

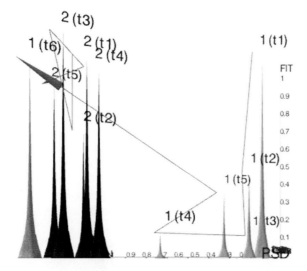

Figure 6.9 The rise and fall and rise of the Feyenoord family while following the municipal executive

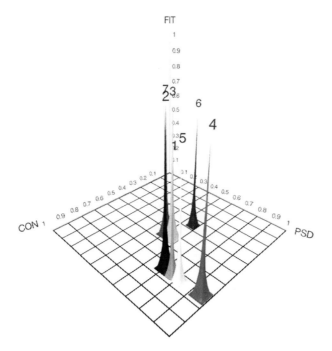

Figure 6.10 Final effort, political disagreement

are clustered, despite the fact that VVD, on the one hand, and GL and SP, on the other, are often said to have incompatible ideas. Indeed, VVD was a member of the governing coalition, while GL and SP were opposition parties. The model enables us to position actors on the landscape on the basis of their true properties. Thus we can develop a more finely grained understanding of how actors relate and why they do certain things beyond a simple centrality score.

We have shown some of the lineage-bound dynamics, with a strong focus on actor movements in search of fitness. For example, the model demonstrates how the movements of actors across two dimensions causes others to move along if they believe that this will return a higher fitness. It also shows that the fields are relatively dynamic. The clusters of actors seem to spread or contract depending on what others do or what external shocks occur at a given point in time. If anything, it reinforces our point of departure that collective decision making shows evolutionary properties. Naturally, other types of questions regarding lineage-bound dynamics can be asked from the model. For now, we would like to discuss the phenomenon of lineages becoming connected.

6.4 COUPLEDNESS

As noted before, lineages can become connected over time. Such connections can have real effects on one or both of the lineages. In some cases, the coupling can even have direct reciprocal qualities, that is, coevolution. There are many ways in which lineages can become coupled, for example through intentional actions from actors seeking to seize opportunities offered by other lineages, or through sheer chance. Our point here is that the coupling of lineages will have an effect on certain properties of those lineages: the events or actions in one lineage could set the boundaries, conditions or space of possibilities for other lineages. Thus couplings provide opportunities for actors to achieve higher fitness if they play well. If they don't, a coupling could spell bad news. Couplings can have the potential to alter the stakes considerably, and we should therefore have a closer look at them. We will do this by studying a case of land use planning in Bangkok, Thailand.

6.4.1 One Step Forward, One Step Back: The Bangkok Case

We shift our attention to Bangkok, the capital of Thailand and one of the world's biggest metropolises. By 2014, it had over 8 million inhabitants (about 12 per cent of the total population of Thailand), and it is still expanding. Growth is a double-edged sword. On the one hand, it is interwoven with more economic activities and benefits such as higher employment and tax revenues. On the other hand, it requires a continuous development of land to facilitate this growth in a more or less controlled manner, which is quite a challenge in such a massive and sprawling city. We will follow two disparate lineages in land use planning in Bangkok. The two lineages became connected with considerable reciprocal effects (see Figure 6.11).

The first lineage concerns the sequences of events related to the redevelopment of a large patch of land east of Bangkok for a new international airport and logistics hub for South-East Asia. Don Mueang had been Thailand's main international airport for over a century. By the 1970s, its capacity was overstretched, and the Thai government decided that a new airport had to be constructed. This decision forms the start of this lineage (field 1). In order to fulfil its dream of a shiny new airport, the government bought a considerable patch of land east of Bangkok called Nong Nguhao. Part swampland, part paddy field, it was drained and prepared for redevelopment. However, civic protests and political instability first, and the Asian financial crisis later, meant that the construction of the airport was postponed time and again.

It was under the first government of Thaksin Shinawatra, who came to office in 2001, that construction was finally commenced in earnest. The motives for building a new airport had expanded since the first plans were hatched. Around the start of the new millennium, other cities in South-East Asia such as Kuala Lumpur had opened new and modern airports. The newly established Thai government believed it would be necessary for Thailand to keep a competitive edge over other nearby countries. An airport was deemed indispensable to survive the economic competition. Thus, the practical wish for more capacity than the old airport could provide was extended to a desire to develop a logistics hub for South-East Asia in order to attract businesses and international travellers from outside Asia. Thus the new airport encompassed not only the construction of the infrastructure associated with airports – terminal buildings, runways, hangars and so on – but also the development of property at the Lat Krabang site directly north of the airport for uses such as office parks, hotels, warehouses and infrastructure related to logistics. The new Suvarnabhumi Airport opened in September 2006. It gave a boost to Lat Krabang and other areas in the direct vicinity of the new airport.

The main actors in lineage 1 are Airports of Thailand PLC, the government of Thailand in its various incarnations, the many property developers and private businesses with a stake in the Lat Krabang area, and the residents of Lat Krabang. The main interest of the airport authorities and

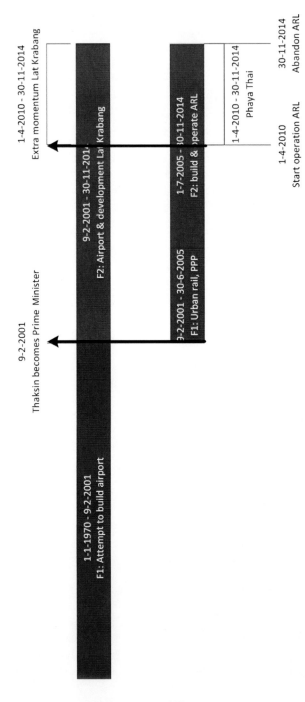

Figure 6.11 Two lineages, two fields per lineage, in the study of land use planning in Bangkok

the Thai government is to establish a vibrant hub that will contribute to Thailand's global competitiveness. The primary interest of the property developers, private businesses and home owners may be a bit more prosaic, namely financial gains. However, it is probably self-evident that these two goals are heavily intertwined.

The second lineage concerns the development of urban rail for public transport in Bangkok. In the 1950s, the city had made the (unfortunate) decision to focus on automobile infrastructure in combating congestion and in promoting economic growth. For a long time, cars were king in Bangkok. The old and relatively restricted rail-bound urban public transport network was completely removed at that time to give way to an exclusive focus on buses as a way to provide public transport and alleviate congestion on the cheap. This didn't work out as expected, and at some point Bangkok became rightly considered one of the world's most congested cities. Subsequently, in the 1970s, there were plans to rebuild an urban railway network. However, progress was slow for various reasons. Probably the most important reason was the regulation that the full construction costs had to be secured ahead of the actual construction of a new line, which was often an impossible step for the cash-strapped government. Railways require a considerable capital investment, and things hardly moved because of that requirement.

As with the case of the new airport, the government of Thaksin Shinawatra accelerated the initiative in 2001. A businessman himself, he decided to expand urban railways in Bangkok aggressively through public–private partnerships (PPPs). Public–private partnerships offer alternative ways of financing and operating public transport. This financial construct would circumvent the perennial shortage of public funds for infrastructure. Consortia of private investors would be offered the opportunity to design, build and maintain such lines in Bangkok. They would be allowed to gain from doing so, for example through ticket revenues or from maintenance fees, as long as they bore the initial investment. There is much potential in such public–private partnerships, but ultimately relatively few initiatives came to fruition in Bangkok (Kakizaki, 2014). However, there was one line that got constructed and put into operation: the Airport Railway Link (ARL).

The decision to switch to PPP in kick-starting urban railways in general and the Airport Railway Link in particular is the start of lineage 2. Here we witness an actor-bound coupling in the shape of Thaksin and his government using the momentum and their own business acumen to convert the many well-intended but never-executed plans into real projects. Although the original plans for the new Suvarnabhumi Airport did not include a new railway connection to the city, Thaksin and his Minister for Transport

Suriya Chungrunguangkit decided that an international airport striving to become a hub couldn't do without such a connection. In September 2003, Minister Suriya ordered a feasibility study. At that point in time, the airport was almost ready for operation, so there was tremendous pressure to get something up and running. Subsequently, the decision to build the ARL was made by the Thai government in June 2004, less than a year after the feasibility study was ordered. Not only is that very quick by Thai standards, but there are actually very few countries in the world that can match that speed of political decision making over capital investments.

In accordance with the new policies, the line was conceived as a turn-key PPP. The consortium would earn its investment back through maintenance fees, revenues from travellers to be collected by the operator, and real estate development in the direct vicinity of the line. The tender was won by Sino-Thai Engineering and Construction Company, with companies such as Siemens and B. Grimm responsible for the electrical equipment. Trains would be provided by Siemens Mobility. After a construction phase marred by setbacks, the line was finally opened as Airport Railway Link in 2010. The operational concept of the ARL centred on a distinction between non-stop express trains for international travellers from the airport to the Makkasan city terminal, and a city train for local people stopping at every station on the line up until Phaya Thai (see Figure 6.12). The main actors in this lineage are the Thai government, the winning consortium and the State Railways of Thailand Electric Train (SRTET, the operator and a new subsidiary of the public State Railways of Thailand).

6.4.2   Reciprocity through Coupledness

We now have two lineages. The first (the development of an international airport and logistics hub at Lat Krabang) went into overdrive because of Thaksin's policies, taking the second (the Airport Railway Link as part of the redevelopment of urban railways in Bangkok) in its wake. The first field in lineage 1 stretches from the decision to buy the land for the airport in the 1970s to the big push from the Thaksin administration in 2001. The actual construction and further development of the area, which extended beyond the airport itself, constitute the second field. Here, decision making concerns primarily what to do where. The Thai government has an influence on that, but one should not forget the property owners, real estate developers, businesses small and large, and the many people already living in the area. They, too, make decisions about if, when and where to invest during the surge of the area. The lineages are shown in Figure 6.11.

The first field in the second lineage starts in 2001, when the Thaksin administration decided to use PPP to kick-start urban rail in Bangkok.

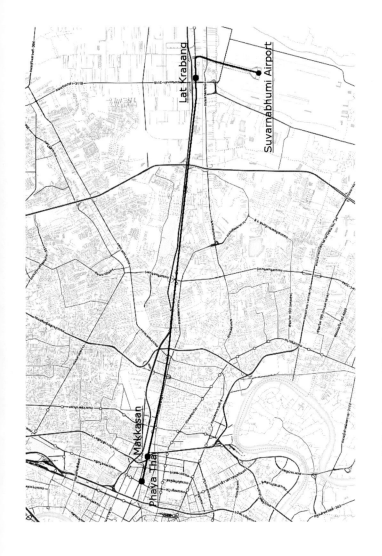

Source: Custom map courtesy of inkatlas.com and OpenStreetMap.

Figure 6.12   *Map of the Bangkok study. The four principal locations, linked by the Airport Railway Link, are highlighted. The express trains would run between Makkasan and the airport without any stops in between. The city train would start at Phaya Thai and stop at each station, including Lat Krabang, before reaching the airport*

161

The first coupling can therefore be attributed to the inauguration of the Thaksin administration, which used the impending opening of the new airport to push the case for urban railways forward by means of a PPP at the same time. This synchronous coupling sets off lineage 2. The first field in that second lineage ends with the signing of the contract with the winning consortium in 2005. The construction and the actual operation of the ARL constitute the second field in this lineage. Arguably, the first coupling was beneficial for the actors in both lineages. For the actors involved in lineage 1, it could enhance the attractiveness of the area because of the possibility of avoiding Bangkok's traffic jams and travelling conveniently between city and airport. For the actors in lineage 2, it meant a breakthrough in the seemingly perpetual stasis surrounding urban railways. The PPP model provided the prospect of at least building one of the many connections needed to improve travel times in the metropolis. The coupling and its reciprocal effects are summarized in Table 6.2.

If the first coupling meant a general improvement for the actors in both lineages, the second coupling results in quite a different outcome. Whereas the first coupling stemmed from an intentional decision, the second takes place through an entirely different, external device. This requires discussing a great number of details, so please bear with us; it will be fun.

Key to understanding the second coupling is a location we did not discuss before and rightly so, because it hadn't played a major role up until that very point. Phaya Thai is one of the main locations in the heart of Bangkok. It provides an interchange between different lines and modes of transport, has a high density of residential and commercial buildings, is close to shopping centres and is generally a pleasant place to stay. The ARL was to connect to Phaya Thai by means of the city train. However, the express trains from the airport would only run as far as Makkasan and terminate there. Consequently, the tracks were designed in such a way that the express trains would not be able to travel on from Makkasan to Phaya Thai. Makkasan was an important site in the PPP model of the ARL. It was situated on a former railway yard owned by the government. This provided a relatively cheap and convenient way of clearing land for the real estate development necessary to recoup the investment costs.

To make all this happen, a big railway terminal was constructed on the site. Makkasan city terminal, as it was named, was going to be much more than just a railway station. Its main feature was that it operated as a remote check-in terminal for the airport. This would enable international travellers to register their flights and luggage at the station and travel to the airport at their own convenience without having to worry about their bags and suitcases, and long queues. In addition, the express line was designed in such a way that it would not stop anywhere between Makkasan and the airport terminal. There was an obvious safety aspect to that, but it was also thought that speed would be important for international travellers. In contrast, the city line train would cover each stop, starting from Phaya Thai in the city centre all the way to the stations at Lat Krabang and the airport terminal. The Makkasan station features a large car park as well as interchange to regional trains and the blue line of the Metropolitan Rapid Transit (MRT) underground trains. All these measures were aimed at attracting international travellers, and the revenues they would bring. In return, it was believed, this would make the Makkasan site attractive for real estate development so that the consortium could earn its money back. In short, the Makkasan concept had to provide the financial foundation for the vitality of the ARL.

While this happened, Lat Krabang kept developing. The allure of the area extended beyond companies looking for a strategically interesting place. For many Bangkokians, Lat Krabang meant relatively affordable housing (because it is not a central location) with a quick connection to both the city centre and the east and south-east of Thailand. As such, it became a blossoming site. In short, the ARL reinforced the development of Lat Krabang. The ARL also provides another source of wealth: land-owners anticipating its arrival had bought up land around the alignment at Lat Krabang and challenged the consortium and the government to build the line. Higher compensation ensured that the project could take place but also meant that the investment sum had gone up.

Now, a number of things happened simultaneously. First of all, journeys on the city line trains skyrocketed owing to the fact that the station of Lat Krabang became well connected to the already-popular Phaya Thai site. If we regard Lat Krabang and Makkasan as communicating vessels, it was Phaya Thai that served as the metaphorical valve that made it very attractive to develop at Lat Krabang (low prices, quick development, good connection to the city centre) instead of at Makkasan (higher prices, still a greenfield site, no direct connection to the city centre). As a result, the development at Makkasan stalled completely.

Second, it turned out that most international passengers also preferred the slower city line train over the express trains, because the city line trains stopped near their hotels, while the express train only stopped at the stag-nating Makkasan site. Journeys on the express trains averaged around 400 passengers per day, which does not make for a sustainable business case. High costs and very few passengers led airline operators to pull out of the city terminal, starting with Bangkok Airways only six months after the terminal was opened. Here we witness a positive feedback loop where lack of travellers decreases the attractiveness of the location, which in turn stalls real estate development and the willingness of operators to staff the

*Figure 6.13   Clockwise from top left: Lat Krabang station with ARL
              at the elevated platform; new city line train; images of the
              deserted Makkasan airport terminal; Phaya Thai station with
              ARL in the background and Bangkok Mass Transit Systems
              (BTS) train approaching the interchange*

terminal, in turn leading to even lower attractiveness for international travellers because the check-in facilities were abandoned.

Third, under-utilization of the express trains led to lower ticket revenues, whilst overcrowding on the city line trains led to shortages of that particular rolling stock and severe problems with maintenance, specifically

Table 6.2   *The first and second coupling and the outcomes in the Bangkok study*

Time	Nature of coupling	Effect on lineage 1	Effect on lineage 2
Field 1	Intentional attempt to reinforce developments	Positive, because airport becomes more attractive	Positive, because urban railway development gets under way
Field 2	Accidental, Phaya Thai and Lat Krabang not anticipated	Positive, because airport and Lat Krabang remain attractive	Negative, because business case for ARL is wiped out

getting enough spare parts to keep everything in shape. In November 2014 it was decided to cancel all express train operations in order to divert the rolling stock to the city line trains. The express trains never returned, and Makkasan lost another reason for its existence.

Linking the bustling Phaya Thai with the attractive Lat Krabang site marginalized Makkasan even before it had started to develop, which is where the business model of the ARL started to unravel. Lack of development at Makkasan meant that a major financial pillar was pulled from underneath the public–private partnership. The synchronous coupling and its effects are shown in Table 6.2. Perhaps Makkasan would have had a better chance if it had functioned as a terminus for all the lines, but it is very likely that that would have made it even less attractive to use the trains. Alternatively, Phaya Thai could have been chosen as the main development site, but that was simply impossible owing to the fact that it was already fully developed. With no alternatives available, the Thai government had to step in and make up the losses.

In short, whereas the first synchronous coupling promoted a positive coevolutionary effect on both lineages, the second synchronous coupling produced a mixed coevolutionary effect, because it meant the further increased attractiveness of Lat Krabang and the airport *at the expense of* Makkasan and the PPP model. In short, the second coupling effectively nullified the space of possibilities for the Airport Railway Link. Naturally, there are all kinds of other possibilities: couplings can emerge through all kinds of dynamics, can be synchronous (as we have seen here) or asynchronous (as shown in the previous chapter), and can cause all kinds of effects. It is up to the research to investigate specific questions, for example about the impact of chance on the fitness distribution in coupled lineages, using the model.

The events described above all took place in 2014, which is where we end our inquiry, but the lineages didn't just stop there. Suvarnabhumi Airport, for all the extra room it provided, quickly reached its own capacity limits. In an interesting rebound effect, the Thai government decided to reopen the old airport for civil aircraft and to encourage all low-cost carriers to operate from there instead of the new airport. But how does one ensure smooth transfers between the two airports? The ARL, which already reached about halfway between the two airports, could now be repurposed as a link between the two. Plans were drafted to extend the tracks of the express train to Phaya Thai and from there on to the old airport. Thus the reopening of the old airport threw a new lifeline to the ARL.

6.5   CONCLUDING REMARKS

We used our fitness landscape model to have a closer look at three main types of dynamics in collective decision-making processes: the search processes within fields as actors explore the space of possibilities for the most promising problem and solution definitions; the adaptive moves of actors across lineages; and the couplings and their reciprocal effects on the space of possibilities between lineages. A recurring theme in these three topics is the evolutionary nature of collective decision making: the space of possibilities evolves over time and in tandem with the adaptive movements of the actors in their attempts to gain fitness. All stock ingredients of evolution have explanatory value here: variation, selection, retention, adaptation, fitness and survival are powerful concepts with which we can understand the dynamics at work. In our view, this strengthens the case for a thorough evolutionary approach of collective decision making. We will use the next chapter to discuss the implications of the evolutionary approach by identifying archetypes of decision making.

7. Evolution in collective decision making

7.1 THE FALLACIES OF CLOCKWORLD AND CLOUDWORLD

It is time for a thought experiment. Imagine you are stranded on an entirely fictitious and deserted island. Everything on this island operates through mechanical systems, so let's call it Clockworld (or *Myst* for game-savvy readers). Pulling a lever will set a mechanism working, for example to provide you with a drink, or pulling another one will get you something to eat. More complicated tasks require you to find the right combinations of levers and switches in order for something to happen – in the same fashion as it takes multiple levers and switches to set a steam engine in motion. Now imagine you would like to escape from the island. Luckily, you have just found a boat, so your escape may be just around the corner. However, the boat is locked away in a shed behind heavy doors, and only the right combination of levers and switches will let you in and the boat out. With no other information available to you, you will need to keep pulling levers and setting switches until you have hit upon the right combination.

This may seem like a complex task, but it is not. It is definitely tedious because of the many possible combinations, but not complex. Clockwork systems always operate predictably: one step forward will always be the same step forward, and likewise one step backward. And that is a good thing, because it allows you to try out different settings without consequences. Nothing is lost when a particular combination doesn't work, because you can simply return the settings to the original state and go back to square one. Your attempt is also helped by the fact that Clockworld is a deserted and isolated place. Apart from you no one else is going to tinker with your settings and you can work in peace. All you need now is time, and probably a drink or two, to get your boat.

Of course, the island in our little thought experiment is entirely fictional and a very unrealistic fiction at that. But it is also a convenient kind of fiction. Since all systems run on mechanisms that behave predictably, we can do all kinds of interesting intellectual exercises to aid you in reaching your goal to get the boat for your escape. For example, we can calculate the number of possible settings you would have to go through towards the solution, or the chance that you will come across the correct combination before you have exhausted all your options, or we can map the series of necessary combinations from starting point to final outcome. We can do all this because we know what you want and we know the order in which you want it, and it is extremely unlikely that you will change your preferences. We can assume that not much else will happen, so we can invoke the ceteris paribus clause for convenience. Once you have found the right combinations for some of the things you would like to achieve, we can help you with predicting the combinations necessary to achieve other goals. In other words, Clockworld presents us with an analytical paradise.

By now, you will probably like to remind us not to get carried away by this little thought experiment. After all, there is very little resemblance between Clockworld and the real world. We concur. Clockworld works exactly *because* it has nothing in common with the real world. Consequently, we need to take the results from our analytical exercise with a pinch of salt. But if Clockworld's properties are unrealistic and therefore of very limited real use, then why is it still arguably the default template for the analysis of collective decision making? Clockworld is not real, but many scholars act as if it is, if not as an outright model then at least by pasting it on to data afterwards to rationalize what people have done and why they have done it – a kind of post-rationalization of the decision-making process. And that is a problem because, while it provides an intellectual stimulus, it tells us little about actual collective decision making in the real world.

Clockworld has a less popular but equally uninformative counterpart called Cloudworld where nothing works straight out of the box, where an attempt to change anything sets off a wholly unpredictable chain of disproportional responses, and where you change your preferences all the time. In many ways, Cloudworld resembles a black box. While the mechanisms of Clockworld allow us to track every change and its consequence, Cloudworld doesn't give us a better understanding of the causal patterns at play. Indeed, it seems that there is no such thing as a causal relationship between your decisions and the outcomes: we only observe our input into the black box and an outcome, if we are lucky. Cloudworld is the realm of randomness, of continuously changing conditions, and (therefore) of unpredictability. Consequently, Cloudworld prevents us from giving precise causal statements, let alone saying something that would amount to a prediction. Its workings obscured, we are left with a constant tinkering in order to let fortune come our way.

While Clockworld's operations are captured in the language of mechanistic causality, Cloudworld's hazy operations require continuous semantic innovation to make sense of them. In a way, that is also convenient. See something you don't understand immediately? Give it a fancy new name. While Clockworld seems the default analytical option for scientists, there are also some scholars who bask in Cloudworld's conceptual inventiveness. Its modest popularity derives from its sense of novelty, because every observed instance seems a new discovery, a continuous reinventing of the world of decision making.

Of course, Clockworld and Cloudworld refer to distinctions between mechanistic and fluid variants of the evolutionary approaches we mentioned in the first chapter. We then promised to offer a third way. To that end, we attempted to unite the elegant but somewhat mechanistic fitness landscape model with the equally fascinating but messy world of collective decision-making processes. Concessions were made. We deviated from the original model in an attempt to make it more suitable for the real social world. However, we also had to cut empirical details in order to make the data processing more manageable. But we hope that, in doing so, we have come closer to the sweet spot where models and empirics combine to create persistent ideas and theories that can be used for further research. The central quest of this chapter is to outline how much more we have come to understand about collective decision making, that is, to map the pathways towards fitness and to explain the nature of those pathways. As Winch (2008) pointed out, showing that something has happened does not constitute an explanation. In getting there, we will walk the thin line between Clockworld and Cloudworld.

7.2   PATHWAYS TO FITNESS

A closer look at the pathways to fitness will give us some first access to the evolutionary nature of collective decision-making processes. Let's start with some basic observations and contrast them to Clockworld and Cloudworld. The most fundamental statement we have made is that collective decision making is a seemingly endless chain of events. Each decision has a predecessor, each outcome a successor. The issue of high-speed railways in the Netherlands has transformed into yet another field at the time of writing this final chapter, as have all the other empirical studies discussed in Chapter 6. It is this continuous unfolding of events that most people will associate with 'evolution', at least intuitively. If anything, it defies the end-game or steady state that is needed for Clockworld to work. That is not to say that actors don't experience something of an end-game, but that is purely because they are myopic and need to have imaginary points in time where the process is 'concluded'. The HSL-Zuid study serves as an example here: it has lasted for almost three decades and counting. No individual has been involved for the entire duration, but the entry and exit of individuals did not start or end the collective decision-making process in any way. It kept unfolding and is still evolving as we speak. By now, NS will have ordered new high-speed trains from Alstom to fully service the line, but needs to combat the gap between now and the moment these trains are ready for operation.

The second observation is that any decision process is heavily influenced by intentional as well as unintentional actions, foreseen as well as unforeseen circumstances. These constitute the events we discussed earlier. In fact, it is most useful to characterize decision-making processes as networks of events. We have attempted to demonstrate this in the HSL-Zuid study by discussing our considerations when connecting events, but the same applies to the more condensed studies presented in Chapter 6. No wonder then that actors face situations they didn't anticipate and that force them to plan, often even improvise, new solutions. No wonder, also, that actors experience their environment as rather volatile. Compare this to Clockworld, where no novelty can be introduced other than in the shape of highly abstract probabilities. But the real world shows that developments seemingly exogenous to the decision matter for the course of the process in all kinds of ways, such as low-cost carriers gobbling up the market share of travellers between European cities or citizens and companies discovering that Lat Krabang is an attractive place to invest in. Perfect foresight would have solved that, but perfect foresight is a figment of the imagination coming from Clockworld. This doesn't mean that we have entered Cloudworld, where randomness rules. As we will demonstrate below, there are certain recurring patterns or sequences of events to be found in the decision-making process.

The third observation follows from the previous two observations. Evolutionary decision making features neither directional law nor necessity; that is, such processes do not move to a fixed end-state. Of course, we can map and explain exactly what has happened. However, we cannot derive statements about the future from that explanation, other than very generic declarations such as that some actors will be more successful than others. Retrospectively, we can even identify certain pathways and can show where path-dependency has emerged. The most prominent example is offered in the HSL-Zuid study, where certain early decisions, such as optimizing the tracks for 300 km/h, kept having an impact on many decisions in the next two decades. Such an effect can also be found in for example the Bangkok study. However, reconstructions such as these do not constitute predictions, because the future is not a perfect mirror of the past, as is assumed in Clockworld. Decision-making processes will keep evolving within a certain bandwidth – that is, they are not entirely random as assumed in Cloudworld – but the role of chance means that new events can set off new directions at any point in time, all of which complies with earlier findings about social evolution as discussed in Chapter 2.

As a result of all this, fitness obtained by actors will be temporal. This should give some kind of solace to those who didn't manage to increase their fitness. Novel combinations of events – in particular unforeseen ones – can set off new directions. For example, the unanticipated accelerated growth at Thailand's new Suvarnabhumi Airport lent a second life to the old Don Muang airport as a makeshift base for low-cost carriers, in turn offering an unexpected second chance to the failing ARL. It demonstrates that the fundamental environmental sensitivity of collective decision making enables opportunities to change course if actors manage to recognize them and seize upon the opportunity. Such opportunities are absent in Clockworld, because the ceteris paribus clause makes it hermetically closed to environmental influences. But then again, it is also not completely Cloudworld, if only because path-dependency severely limits the number of possible outcomes once things have been set into motion. Decisions made in one field and lineage can, and often will, set the conditions for subsequent fields or other coupled lineages.

How do these three fundamental observations relate to the actors in the various fitness fields? We have seen how decisions turned into actions and interactions with actual consequences that set the conditions for the next field. There is nothing much mechanical about it. Actors may find themselves in a classic negotiation game – recall the staring contest between NS and the Minister in the HSL-Zuid case – but it is equally likely that they will muddle, improvise and stumble their way from one field to another. And so we can conclude, in general terms, that the decision-making process keeps unfolding and meandering within a certain bandwidth. Each situation is responded to through action and interaction on the basis of information and preferences prevalent at that particular point in time. This, of course, includes prior knowledge and preferences. Fitness, then, is obtained as a result of one's own actions, formulated in problem and solution definitions and connections, in conjunction with favourable circumstances.

7.3 VISTAS OF LANDSCAPES

The observation that fitness is reached as a result of the interaction between an actor's capacities and actions and the circumstances in which it finds itself draws our attention to the landscape, that is, the surface upon which actors move (see Gavrilets, 2010 for a discussion of various types of landscape). The idea that actors 'walk' on a mountain-like landscape, a dynamic landscape in some accounts, seems to resonate strongly with certain researchers looking at collective decision-making processes. However, a landscape in the sense of a surface plot belongs to Cloudworld. It presents a nice semantic innovation and could serve as a heuristic device but has little to do with the social reality of decision making. The only way to make it work would be to have a data point for each possible coordinate on the grid, and to assign fitness to each data point. Suppose that we have found one actor i (PSD$_i$ 0.4; c_score_i 0.4; f_i 0.8) and another actor j (PSD$_j$ 0.6; c_score_j 0.6; f_j 0.6) in an empirical data set. That would give us just two data points in the field but nothing else. A continuous surface would require data points for *all* the other coordinates. Since these are absent in our – or any, for that matter – empirical data set, we would have to make them up. In other words, we would need to create a third imaginary actor k (PSD$_k$ 0.5; c_score_k 0.5; f_k 0.7), and another actor l, and so on. Next, we would need to assume (i.e. simulate) that all actors are distributed evenly across all the coordinates, and we would need to assign fitness to each actor – which, again, would be nothing but a guess. The empirical research in this book has shown that in reality it is much more likely that actors either cluster around the same or very similar coordinates or are loosely scattered across the field, and that fitness comes from the interaction between actors. In short, a continuous surface is impossible when working with real-world data – even a cursory glance at the findings presented in the previous studies will confirm this – and can only be achieved in a simulation, which, while interesting, violates our prerequisites.

The alternative, then, is the more realistic version presented in this book. The final model, and the ways in which we used it for the analysis of collective decision-making processes, features components from the original model as well as from network theories and methods. In particular the interaction component – expressed here in the c_score – bears a resemblance to social network analysis (SNA). However, it adds the substantive component (PSD) and fitness to that network, which is a novelty in SNA. The model seems to be more of the network type than a 'true' fitness landscape model, and we therefore refer to it as a fitness *field* instead of a landscape. Yet this doesn't mean that the landscape has disappeared. We assigned fitness retrospectively and can therefore point to regions as configurations of PSD and c_score where fitness is more likely to occur. These are not hills or slopes to be climbed, or valleys to be avoided, but they are combinations that are more likely to lead to fitness gains, and we concur that certain actor types will have more difficulties in getting into these regions than others. And so that is what we will focus on in this chapter.

7.4 ARCHETYPES AND RULES OF THUMB

In the following, we will present six archetypes of collective decision making: three concerning actors vis-à-vis other actors, and three concerning interaction. The purpose of these archetypes is to flesh out the dynamics described above in more detail and with more clarity. Archetypes summarize and represent our empirical findings in a somewhat purer form. As such, they serve as empirically founded abstractions that can be used for further exploration, research and even simulation in order to tease out the precise mechanisms that govern these particular archetypes and to see whether they emerge in other studies. To that end, we summarize each archetype in a so-called rule of thumb. A rule of thumb constitutes a pattern that describes data that couldn't be described more succinctly by just listing that particular data (cf. Dennett, 1991). It is a principle that seems to hold well for the studies carried out for this book, but it should not be regarded as a universal law. Rather, such rules are to be seen as testable and grounded hypotheses that invite further research. In this exercise we will again walk the thin line between Clockworld, where every rule has universal value, and Cloudworld, where no rules can be discerned.

7.5 ACTOR ARCHETYPES

The actor archetypes show consistent behaviour in the respective studies. This consistency is not just a trait for the actor but also affects the space of possibilities and/or behaviours of other actors. There are three main archetypes with many minor variations within. The first archetype is that of the buoy: an actor that is relatively stable and that points the way for other actors. The second archetype is the jumper: an actor that behaves relatively erratically and can be seen to be moving around on the grid. The third archetype is the inflexible, which is an actor that against all the odds (un)knowingly or (un)willingly sticks to certain elements of its PSD.

7.5.1 The Buoy

The buoy represents actors that are relatively stable when it comes to their position on the fitness field, and that provide a (proverbial) beacon for other actors to use for navigation during the decision-making process. Its relative stability holds that it *can* alter or change its position in the field, but these changes are mostly minor. In other words, the buoy occupies the same or a similar position throughout. We found that the buoy is by and large publicly engaged (think of a governmental body with a broad set of tasks) and channels all kinds of (societal) pressures into the decision-making process. This means that the buoy will feature many and diverse elements in its PSD. By definition, many of these elements will be similar to those of other actors. However, the centrality of the buoy is not only due to it having many elements in its PSD; it is also due to it being connected to most, if not all, the other actors in the field. This high centrality means that many actors (un)knowingly influence the amount and diversity of elements in the buoy's PSD, as the buoy mutates elements or picks up elements from other actors in the field and integrates them into its own PSD. This is a reinforcing mechanism, as it is expected by the actors that the buoy will pick up these elements, and the buoy expects that the other actors will expect that the buoy will do so (cf. conventions of Lewis, 2008; Llewelyn and Lewis, 1970). In other words, this uniform conformity reconfirms the buoy's central position, that is, an expansion of the number of elements in its PSD, including the shared ones, raising its *c_score* once more.

The central position doesn't concern the current decision-making process exclusively. It also happens in analogous situations because of widely held expectations that in new situations it will do so as well (cf. Sugden, 1986: 50). In other words, other actors focus on the buoy to help them find their way in new uncharted territory. Owing to the large number of (diverse) elements in the PSD and the public role, the buoy has to make relatively many compromises. Thus it is very hard for the buoy to succeed in realizing *all* the elements of its PSD all the time; hence the buoy will more likely realize a certain number of elements of its PSD. In short, the buoy will differ in attributed fitness values (there is some resemblance with a real buoy and the tides it has to withstand).

Actors learn from past experience (see the relation with the starting input for the *c_score*) that playing a certain role, making connections with or refraining from participation with others, or adopting a certain strategy will be successful in order to achieve certain problem and solution definitions. In evolutionary terms, the successful mutant strategy is selected and adopted, which becomes the shared unique behavioural rule. Thus certain conventions start to evolve (Maynard Smith, 1982; Schelling, 1960; Vega-Redondo, 1996). The prominence of the buoy as guiding device is such a convention. The buoy archetype does not have to be present in any collective decision-making process, but will most likely be present in a *public* decision-making processes. Be it a collective or public decision-making process, as a rule of thumb the behaviour of the buoy as well as the other actors in the configuration can be typified as shown in Box 7.1.

The position of a buoy will by and large look like that in Figure 7.1. As is obvious from the figure, the buoy is able to move around, but sticks in

BOX 7.1

An actor is a buoy when it has a PSD between 0.7 and 1, and a *c_score* between 0.7 and 1, and all the other actors in the configuration have PSDs and *c_scores* lower than those of the buoy. The actor behaves as a buoy (B) because all actors in a recurrent situation, that is, within the lineage or between similar lineages, expect that, and it is common knowledge in the configuration that:

1. everyone conforms to B;
2. everyone expects everyone else to conform to B;
3. everyone prefers to conform to B on condition that the others do.

This results in a fitness value for the buoy that ranges between 0 and 1, with the median around 0.75 in a left-skewed distribution.

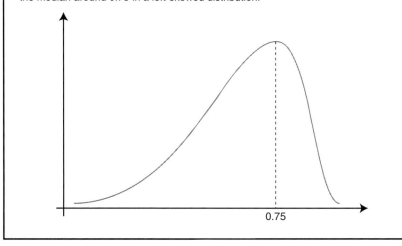

0.75

the corner with the high connection score and many (similar) elements in its PSD.

The Minister of Transport in lineage 3 of the HSL-Zuid study serves as an example par excellence of this archetype (see also Figure 5.20). The Minister owns the infrastructure and initiates a tendering process for the concession to operate. She can therefore set the demands. However, she also has to react to every minor or major incident during the process because of her very public role and the pressures exerted on it, for example by the Parliament. She incorporates ideas from others and thus widens the space of possibilities. Other actors need the information and permission of the Minister and are connected or become more connected to the Minister as a result of similarity in PSD elements.

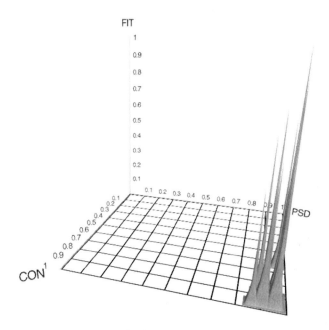

Figure 7.1   Representation of a buoy

Do keep in mind that an actor may be a buoy for several fields in the lineage but may change its position as a result of the issue at hand, possibly putting another actor in the position of buoy. Owing to her properties such as tasks assigned, budget and legal powers, the Minister of Transport in the HSL-Zuid study is practically always the buoy no matter which lineage is studied. However, she does swap position with NS when NS starts the public tendering process for high-speed trains. The same type of swap happens with the municipal executive and the Feyenoord family in the Sports in the City study. In other words, the change from public decision making to public tendering changes the respective roles of the publicly engaged actor and the private party. The private party in the tendering process of course functions as the buoy, because all the other actors focus on its demands with regard to the tender; that is, all are connected to the buoy and will follow or inform parts of the PSD of the buoy.

Centrality and decision-making power don't mean that buoys can push anything through. They are often unable to attain and retain a constantly high fitness. However, they are able to exert considerable power, as many other actors depend on them. They can try to get things their way by influencing other actors' PSDs, that is, act as selection mechanisms for others, such as the Bundesrat did in the Gotthard study. For certain actors the buoy functions as an anchor point; they focus on the buoy because that focus offers an opportunity for success. Reciprocity may form the basis for the behaviour that follows: that is, if we support the buoy, the buoy will probably or hopefully support us. These followers can become jumpers when they see the opportunity to raise their fitness by following different actors they think will help them to succeed.

7.5.2 The Jumper

The jumper represents those actors that are relatively volatile in the landscape. In general, the jumper behaves opportunistically in search of (short-term) fitness gains. This propensity towards opportunism is shown in considerable changes in the elements of the jumper's PSD when it searches for those combinations that promise a better chance of obtaining higher fitness. This happens because the jumper sees elements of the PSDs of others that are similar to elements of its own PSD and therefore tries to align with those actors – the caveat being that those other actors are generally seen as successful. In the process of alignment, the jumper raises its *c_score* as well as the *c_score* of others. Conversely, the jumper can opt to take an opposite position to that of certain other actors, as the jumper sees elements in the PSDs of other actors that contradict its own position. The jumper will take the opposite position and contra-align when it believes that this is the more successful thing to do. Thus the jumper is not just a follower. In the case of taking an opposing position the *c_score* of the jumper will be lowered and possibly that of other actors as well. Arguably, the actual success rate is not always what the jumper expects; sometimes it works to align or oppose, but sometimes it creates a situation in which fitness gains are unlikely. However, it can be speculated that sometimes actors take an opposition position to make sure that certain unwanted elements in the PSDs of other actors (from the jumper's perspective, of course) are *not* realized. We believe that this expectation is reasonable, but it is speculation, because this kind of data is hard to get.

The volatility of the jumpers introduces dynamics into the landscapes as actors jump around and others try to maintain a relationship or try to avoid them. As a rule of thumb the behaviour of the jumper can be typified as shown in Box 7.2.

By and large many political parties will show the behaviour associated with jumpers. An example can be found in the Sports in the City study. The political party GL is a coalition member in the earlier stages (i.e. field 1 and the beginning of field 2) of the decision-making process, but moves to the opposition after the municipal election in 2010 (i.e. fields 2 to 5). Initially it is in line with the municipal executive, that is, the buoy, in this

BOX 7.2

An actor is a jumper when it has a PSD between 0.1 and 1, and a *c_score* between 0 and 1. The actor behaves as a jumper (J) because it (dis)connects with other actors, as it sees opportunities to achieve elements of its PSD, resulting in changing values of its PSD and *c_score* throughout the lineage, which is common knowledge in the configuration. The jumper has fitness values ranging between 0 and 1, with the median around 0.5 in a normal distribution.

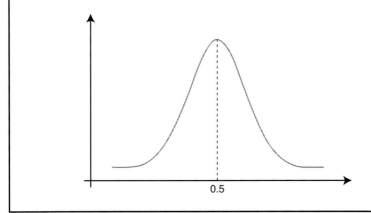

0.5

process, but it shows its true politically left-wing colours by making clear that if there were really a wish for a referendum it should be binding rather than just informing. With this stance, the party opposes the general idea of most other parties and the municipal executive. In the third and fourth fields it plays hardly any role and mostly follows the ideas of the municipal executive or the Feyenoord family. However, it shows two faces in the final field. It tries to follow the municipal executive but at the same time adhere to the pressures from major media attention, based on the emotions of the Feyenoord fans, that the Save De Kuip plan should be taken seriously. This is visualized in Figure 7.2.

Volatility has its advantages and drawbacks. It is worth noting that jumpers move between positions with regard to the specific issue being researched. It does not mean they change all their fundamental properties all the time. The political party discussed above may have changed its ideas about the new stadium and may have sought alignment with certain actors, but it did not stop caring about the natural environment – one of the main characteristics from which this party draws its popular vote. The jumper represents the somewhat erratic movement of actors across the landscape.

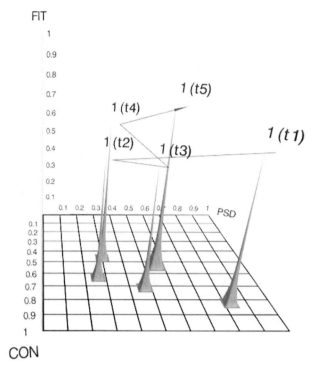

*Figure 7.2   GL jumping around, sometimes sticking to other actors,
sometimes doing the opposite of what other actors do*

7.5.3   The Inflexible

Where volatile actors are represented through the jumper archetype, inert
actors are represented by the inflexible archetype. This archetype concerns
actors that attempt to stick to their position no matter what the situation.
Note that this is rather different from the buoy. The buoy is relatively open
to what happens around it, which explains its high *c_score* and diverse elements in its PSD. The result is some movement. The inflexible, however,
wants to stick to certain or all of the elements in its PSD. This means that
the inflexible may stay put even when that is no longer a viable option.
Delays in the selection mechanism lead to retention of (elements of) the
PSD of the inflexible that won't be selected in the long run. There are
two reasons why an inflexible will (un)knowingly stick to its problem and
solution definitions. First of all, it results from a myopic overview of the
configuration. In other words, the inflexible has lost touch with the general

movement and clustering of other actors in the configuration in terms of both *c_score* and PSD. There can be many reasons for that, such as group think, autopoiesis, or the lack of capacity to process environmental information. The second reason concerns a normative understanding that one needs to stick to one's ideas and ideals, which exclusively concerns problem and solution definitions. In any case, the chances of the inflexible achieving its goals are slim, because the environment – including, of course, the complete configuration of actors with their PSDs and connections – selects for a different problem and solution definition.

It is hard to visualize the inflexible in archetypical fashion for two reasons. First, the inflexible holds on to one or more elements in its PSD, but the totality of elements may change. This means that on certain elements the inflexible can connect to other actors in the configuration, but at the same time persistently holds on to certain elements that no other actor shares. The position of the inflexible can thus change in the field whilst it stays inflexible. Second, and more importantly, the rest of the configuration changes, owing to actors changing elements in their PSDs and *c_score*. This means that the field's dimensions will change during the transition from one field to another. The changing field makes it difficult to pinpoint the inflexible even if it hasn't adapted new elements in its PSD because the visible position of the inflexible on the grid is relative to others. In order to pinpoint an inflexible actor the number of times an actor holds on to certain problem and/or solution definitions over the different fields in a lineage, that is, retention, needs to be calculated and analysed in comparison to the other actors that do not adopt or abandon those specific definitions.

Notwithstanding the impossibility of visualizing the inflexible, as a rule of thumb the behaviour of the inflexible can be typified as shown in Box 7.3.

An example of the inflexible can be found in the shape of NS and the Minister of Transport in the HSL-Zuid study during the tendering of the concession. NS was convinced it didn't really need to compete for the concession to operate in a public tender because it simply couldn't believe that the Minister would really consider the possibility of a new operator on the Dutch network. It thought that the Minister was bluffing, even when the Minister fired a warning shot by turning down one of the early offers NS made. This resulted in a showdown, ultimately causing the Minister to open a public tender and NS overstretching itself financially in a desperate attempt to regain the ground lost. In fact, the Minister was also not immune to inflexibility. One of the most persistent elements in her PSD was the requirement for a maximum speed of at least 250 km/h, even though other actors had already demonstrated that this was not very

BOX 7.3

The inflexible (I) can have a changing number of elements in its PSD, but will stick to certain elements, which means it has a PSD between 0.1 and 1. It is common knowledge in the configuration that (a) the inflexible will disconnect when holding on to its own inflexible PSD elements, and (b) it will connect through its non-inflexible PSD elements. 'A' will lower the *c_score*, and 'B' will raise the *c_score*. The inflexible has a *c_score* between 0 and 1, and will hardly move in the configuration. The resulting fitness value for the inflexible ranges between 0 and 1, with the median around 0.35 in a right-skewed distribution.

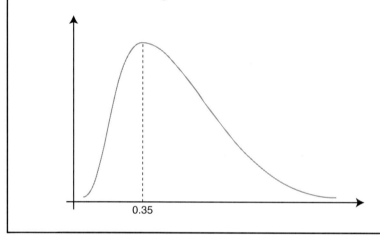

0.35

efficient from an operational point of view. This stubbornness in the face of changing circumstances can be traced all the way back to the very early decision that the tracks would have to be suitable for such speeds in order not to be locked in if higher top speeds became economically feasible at some time in the future. What was supposed to be a decision that would give extra degrees of freedom in the future turned into a requirement cast in concrete.

On the whole inflexibility seems to return unfavourable results, that is, low fitness. Why then do actors sometimes cling on to something unattainable? We have already referred to aspects such as stubbornness and group think. However, we would like to point out the fact that sometimes the landscape just evolves in a certain way unbeknownst to the actor in question. It could very well be that NS was actually correct in thinking that competition in the Dutch railway sector would be frowned upon in general, but then overlooked the fact that competition on the HSL had become a very real possibility as a result of the changing positions of other actors.

In other words, inflexibility can happen knowingly as well as unknowingly, willingly as well as unwillingly. But, whatever these considerations, the pay-off is unfavourable.

7.5.4   Actor Archetypes and Fitness

The obvious question now is: is there a clear relationship between certain actor movements, actor properties, and fitness across the archetypes? The answer is: yes, but not in an absolute sense, or at least not in the mechanistic terms of Clockworld. As mentioned before, the buoy has a considerable chance of fulfilling at least some of the elements of its problem and solution definitions on account of having a very varied set of goals – at least some of them will be reached. Naturally, actors that manage to align with the buoy in terms of both c_score and PSDs may also increase their chances of success. This is particularly the case when actors hold a simple PSD as a subset of the buoy's larger set of elements in its PSD. If the buoy manages to attain those specific elements, the other aligned actors having that subset will follow accordingly.

Having said that, it would be too simplistic to say that clustering is the only way to attain fitness. If we take the many tendering processes in the HSL-Zuid as an example, we can see that there is considerable distance between the tenderer and the offering parties, where it is the offering parties that cluster owing to their similar connections and offers (in order to meet the tenderer's requirements) but (obviously) not with the tenderer. Yet it is very likely that the tenderer will get what it wants, as well as one of the offering parties. In short, it is more than likely that an actor will gain higher fitness when clustering, that is, raising the c_score, but this is neither sufficient nor necessary. An actor – especially one in a tendering process – can attain fitness with a low c_score but needs a number of shared elements in the PSD. That is, having a high(er) number of shared problem and solution definitions is a necessary condition. Or, to put it differently, alignment of elements in PSDs seems to be relatively more important than nurturing many connections – keeping in mind the configurational nature of connections and substantive arguments, of course. However, we have also observed some instances where an actor took the conscious decision not to align in order to prevent another actor from realizing (parts of) its PSD. For example, the political party LR in the Sports in the City study announced it would vote against the new stadium a day before the actual ballot, leading the municipal executive to withdraw the plan from the agenda in order to avoid further complications. We should also point out that not every actor has the ability to cluster. Basic resources, such as money, legal power and knowledge, enable or restrain actors. So, yes, there are other likely ways in which one can achieve, but these are not mechanistic recipes for success.

7.6 INTERACTIONAL ARCHETYPES

Having discussed how actors and their characteristics interact with other actors and their characteristics and how that produces fitness, we now shift our attention to three archetypes of interaction. These archetypes concern typical dynamics that occur in the interaction and their results in terms of fitness. The first archetype is that of attempts to break out of deadlocks through changes in the c_score. We call this 'force to fit'. The second archetype concerns the idea that some of the clustering due to alterations in problem and solution definitions, as expressed in the c_score, is involuntary, that is, self-organized entrapment. The third archetype centres on the idea that the introduction of diversity in the c_score and problem and solution definitions triggers more diversity; that is, diversity breeds diversity. The interactional archetypes can't really be suitably visualized, so we will describe them in detail instead.

7.6.1 Force to Fit

It may happen that the configuration of actors gets stuck in a certain situation from which quick fitness gains for any of the actors in the configuration seem impossible – in fact, 'may' could be quite an understatement. Increase of interaction by establishing more connections or actively creating more similar problem and solution definitions, i.e. raising the c_score, has the potential to alleviate the deadlock. We call these intentional actions to escape such dead ends through an increase of connections 'force to fit'. There are three ways to force fit.

The first strategy is by actively building connections with actors in other lineages in order to achieve more successful combinations. That is, actors consciously establish couplings between lineages in order to get momentum in one or more of the coupled lineages. The Bangkok study provides an example of that. Prime Minister Thaksin's entrepreneurial approach to developing infrastructure in the shape of public–private partnerships connected the development of the new airport and surrounding area with the lacklustre attempts to get an urban railway system up and running. Both initiatives benefited tremendously from this coupling, raising the fitness score for all the actors involved – in the short run, that is. The effects in the long run were less positive for the public–private partnership in the railway project, so the results of the strategy to couple lineages may vary.

The second way to force fit seems similar to the first one but differs in one important aspect: the coupling is established with a collective decision-making process that is not in any way connected to the first one. The connection in Bangkok between the airport and the Airport Railway Link was obvious because of the shared physical location, even though it required a new Prime Minister to really bring them together in the planning process. But one can also bring an unrelated issue to the table if that affects the actors one is trying to hammer out a deal with. For instance, if actors hold diametrically opposed problem and solution definitions, they may seek to incorporate the problem and solution definitions of a completely different decision-making process to create a package deal. The package deal will raise the number of similar elements in the PSD, subsequently raising the *c_score*. This may then break the deadlock and raise the fitness of the actors involved. An example of that can be found in the HSL-Zuid study when the Dutch government decided that the deadlock between them and the Belgian government could only be lifted by adding the Westerschelde issue to the equation. There is no real relationship between the construction of the HSL and the deepening of the Westerschelde, the first being about high-speed trains and the second about shipping and nature development in a geographically different area. But, since the Belgians had some clear preferences in the Westerschelde issue and the Dutch in the HSL issue, it was only a matter of time before the two issues were linked in order to force fit. Arguably, in hindsight this was a somewhat successful move, as all the actors got a bit of what they wanted but had to make concessions too.

The third strategy to force fit is to connect with other actors that were previously uninvolved and that don't share anything in terms of PSDs. Such a strategy sees actors propose certain problem or solution definitions that are connected to actors exogenous to the process. Those actors are deemed useful in breaking the deadlock because of their resources or capacities. Their involvement raises the number of connections as well as the number of similar elements in PSDs and thus the chances of realizing certain problem or solution definitions. However, the exogenous actor may also stay that way and not become involved. If that happens, the deadlock will most likely not be lifted, as that specific solution definition is doomed to fail. For example, the municipal executive of Rotterdam tried to get the city region authorities involved in organizing funds for linking the existing metro network of Rotterdam with the new Stadium Park location. The city region was an outsider to the decision-making process, because its duty is just to manage public transport. It also didn't plan on building such a metro line, regardless of Stadium Park. Still, the municipal executive tried to incorporate it in the decision-making process as a way of enlarging the

---

BOX 7.4

In the event of a deadlock caused by enduring opposing PSDs or long-term inertia on both *c_score* and PSD, actors will actively attempt to establish coupling outside the current lineage. Consequently, environmental conditions, external problem and solution definitions and/or external connections are imported into the current lineage to raise the possibility of realizing (more) elements in the PSDs. Successful force to fit raises the *c_score* and fitness.

---

proverbial pie. As we have shown, the attempt was unsuccessful and the metro was shelved.

Force to fit, then, constitutes all attempts to remove deadlocks and enable fitness gains by trying to raise the *c_score*, either by establishing new connections or by creating more similar problem and solution definitions. Even though there are three subvariants, a rule of thumb for force to fit can be typified as shown in Box 7.4.

The results in terms of fitness gains are decidedly mixed – some gain, some will not – so it is not possible to establish an unequivocal relationship between coupling and (changes in) fitness. It seems plausible that changes in the *c_score* can lead to improved fitness in cases where all the actors have got stuck. Such findings have been presented elsewhere in the literature, but we would like to remind you that, in the long run, no actor is able to reach *and maintain* a continuous high fitness, including after coupling has been established. Notwithstanding the outcomes, the archetype represents the idea that enforcement can create a way out of deadlocks.

7.6.2 Self-Organized Entrapment

In the configuration actors can have connections with certain other actors but not with all, and this holds for all sorts of actors. If certain problem and solution definitions of an actor seem promising other actors are likely to incorporate those definitions because they may bring higher fitness. Especially if the actor that holds the promising PSD is a central actor, for example a buoy but not necessarily always a buoy, it will cause some of the other actors to incorporate the promising definitions. Thus these actors will get connected or get more closely connected to the central actor. In the event that a non-central actor holds promising definitions, other actors may still incorporate these definitions too. It is then very likely that certain central actors will incorporate such promising definitions, which means that those central actors will expand their problem and solution definitions and/or make new connections. Actors connected to this one will move along involuntarily owing to the interdependencies between them. The movement can be considered as superimposed by the leading actor, even if this actor didn't necessarily intend to take other actors along, that is, self-organized entrapment.

If any actor with a high c_score and many elements in its PSD expands its problem and solution definitions or makes new connections, it can change the configuration because of these interdependencies. That is, the alteration in PSD and/or c_score by the central actor causes alteration in the c_score of the connected actors. Therefore the whole configuration changes, but the central actor remains central. Self-organized entrapment can be traced back to the distribution of connections across all actors in the landscape. In most cases, the total number of connections is distributed asymmetrically, with certain actors being more networked than others, such as the buoy. Actors that are very well connected are more likely to cause this entrapment. We didn't observe any clear fitness gains for all the actors involved but also no clear disadvantages. Given the large number of elements in the PSDs, the outcomes for all actors are more likely to be mixed: some will gain, and some will lose. As this is a self-organizing process, we should be careful not to see it as a conscious strategy to gain fitness. Although the results may be mixed, a rule of thumb for self-organized entrapment can be typified as shown in Box 7.5.

An example of self-organized entrapment can be found in the interaction between the Feyenoord family and the municipal executive in the Sports in the City study. In the process, they depended heavily on each other for their resources, but the Feyenoord family depended more on the municipal executive than the other way around. Naturally, the municipal executive would regret a failure to develop a new stadium, but it would survive, because its PSD is much more diverse than just building a new stadium. For the Feyenoord family, the stadium is essential to its (financial) survival. It has the advantage that it owns the current stadium and can count on a large fan base to fill the new one, but it has the disadvantage that it doesn't have enough financial clout to pull it off on its own. The presence of the

---

BOX 7.5

If throughout a lineage an actor with a high c_score and many problem and solution definitions is present and the other actors are scattered across the configuration, it is likely that the latter will follow the central actor (in)voluntarily as it changes its PSDs and/or connections, resulting in varying fitness gains and losses.

Feyenoord family is a necessary but not sufficient factor in getting the new stadium, but the presence of the municipal executive is both necessary and sufficient for the stadium to be constructed. That renders the Feyenoord family more dependent on the municipal executive. Consequently, this actor had no choice but to follow the municipal executive, even though the latter didn't set out to lead the Feyenoord family. Originally, the municipal executive had wanted to develop the stadium as part of a larger sports campus project. When this failed, the municipal authorities urged the Feyenoord family to make plans to develop a new stadium on their own. When that attempt failed, the municipal executive urged a public contest, even presenting Save De Kuip as a viable alternative. And, when that attempt also failed, the municipal executive proposed going back to the first option. Other actors, most prominently the Feyenoord family, moved along despite their own ideas. However, it would not do justice to the situation to say that the executive forced them to change all the time. Behind the scenes there was considerable communication and coordination between the two. Therefore, it is safe to say that the Feyenoord family at the very least complied with the proposed changes. The movement wasn't entirely forced but was also not entirely voluntary. Given this observation, and the fact that some but not all actors moved along, we can speak of a process of self-organized entrapment.

7.6.3 Diversity Breeds Diversity

The third archetype stems from the observation that the introduction of diversity in the configuration, in either the *c_score* or PSD, or both, has a tendency to create more diversity: diversity breeds diversity. So how does this come about? In decision making on a certain issue it may be that factors such as media coverage cause actors previously not involved to think that they need to be involved, or actors that are already involved want others to get involved too. The voluntary involvement of other actors broadens the selection mechanism. Note that this is not just about connecting with more actors but about connections with more *diverse* actors, raising the number of similar problem and solution definitions and related *c_score* for many actors in the configuration. Note the contrast with the force to fit archetype, where an attempt is made to import non-involved actors or their problem and solution definitions into the lineage in order to break through the deadlock.

Diversity may also grow as a result of actors getting informed by the way other involved actors look at the problem. That is, the fact that actors have certain problem or solution definitions that another actor couldn't imagine beforehand now makes that actor formulate new related definitions, expanding the variety of elements in its PSD. The expansion of relatively similar definitions means a rise in c_score, and vice versa. Introducing more substantive diversity can lead actors to consider new connections with actors they were previously unconnected with, causing another increase of the c_score. In return, connecting with more diverse actors can lead to substantive changes, as seen in the expansion of the PSDs. Put in simple terms, exposure to diversity leads actors to see what there is on offer beyond their own myopic view. There is a point during the process where this diversity will no longer increase. This could be a natural saturation point, but it is more likely that interactions between actors (such as in the shape of a pending decision) as well as environmental conditions (such as legal ruling imposed by a court) will lead to certain PSDs being selected. We have demonstrated that such points will be reached during the process, but we can't predefine the exact moment when this will happen; that is, it is a context-specific limit.

As with any feedback loop, there is no obvious starting point. However, the principle boils down to the fact that exposure to diversity in problem and solution definitions and connections can lead actors to rethink their PSD and connections. We therefore use the term 'diversity' instead of 'variety', because the latter concerns PSDs only. Does the introduction of more diversity lead to higher fitness? Maybe it does. The archetype implies that connecting with more diverse actors, including actors that had no prior connections with each other, or incorporating more diverse problem and solution definitions raises the likelihood that elements of the PSD will be fulfilled. In certain cases, however, it may also be that diversity means making more compromises because there are more connections with more actors and/or more diverse elements in PSDs, which makes it harder to fulfil all wishes. Thus fitness can't be attributed conclusively. The rule of thumb for diversity breeding diversity is as shown in Box 7.6.

The prime example of how diversity breeds diversity is the process we described in the Gotthard study. The space of possibilities expanded as many actors in the region wanted to be involved in the decision making

---

BOX 7.6

If actors in a decision-making process hold diverse PSDs and if they are connected to diverse actors, it is likely that the number and diversity of actors, the PSDs and the c_scores of the configuration will rise, which will set off a self-reinforcing effect in which the diversity brings forth more diversity until actor interactions or environmental conditions select certain PSDs, which may, but not by definition, lead to higher fitness for certain actors.

on the basis that all of them suffered from the issues in that region. As more diverse actors brought more diverse perspectives to the table, the exploration quickly moved beyond Porta Alpina. Actors discovered, first, that there were indeed other actors they had not worked with before and with which alignment could increase fitness and, second, that the space of possibilities was much richer than the original PSD around Porta Alpina. Certainly, the second point didn't just mean that more solution options were available but also that more problem definitions were formulated. The increase of diversity came to a halt as a result of time pressure imposed by the deadline to make decisions. Subsequently, actors refocused on their own interests and the diversity diminished somewhat.

The Gotthard study was inconclusive about whether the introduction of more diversity led to fitness gains or losses. It is true that the actors had developed a better understanding of the situation they were in, and that they had identified promising solutions for that situation. However, the implementation of the plan would take many years, and there was no immediate indication whether the solution chosen would return favourable results. If we are a little less cautious, it sounds plausible that more diversity will lead to higher fitness, simply because it allows actors to explore more options and therefore get a better understanding of how to achieve fitness. At the same time it means making more compromises – as we have outlined before. We have to be content that our findings are inconclusive at this point. However, the archetype is further consolidated when we consider its opposite, where lack of substantive diversity means stagnation. This is exemplified in the Sports in the City study, where the pool of problem definitions showed little initial variety and little change in that variety across the fields, and where the pool of solution definitions demonstrated how all the solutions were framed from within the actors' organizational boundaries and geared towards that very limited set of problem definitions. This counterfactual finding bolsters the archetype that diversity does indeed breed diversity.

7.6.4 Interactional Archetypes and Fitness

The interaction archetypes represent particular dynamics that result from the conjunction of actor properties, their strategies and environmental factors. Again, the question is whether we can assign an overall and unambiguous relationship between these interactions and fitness. The question is much harder to answer here than in the case of the actor archetypes for two reasons. First, the archetypes all represent interactions between multiple, sometimes all, actors in the field. Some will gain and some will lose fitness in the process. Force to fit may be the archetype coming closest to what could be called a strategy, but here we see – as with the other two archetypes – that the results are mixed, as compromises have to be made. Second, the archetypes represent the dynamics of interaction, and that is much broader than just the deployment of strategies and subsequent responses. 'Self-organized entrapment' and 'diversity breeds diversity', in particular, constitute interactions that are relatively independent of intentional strategies (though such strategies can be part of the interaction, of course) and of the individual characteristics of actors (though such characteristics matter too).

Even though the interactional archetypes do not feature an unambiguous relationship to fitness, we believe that they offer pertinent insight into the evolutionary nature of collective decision making. They demonstrate the effects of changes in the space of possibilities and subsequent deployment of selection mechanisms on the decision-making process (force to fit, diversity breeds diversity), how certain choices and actions can bring forth self-propagating dynamics that start to have a life of their own (self-organized entrapment, diversity breeds diversity) and consequently how the process can evolve in such a way that actors will be deprived of their autonomy and pushed into reactive roles (something that can be observed in all three archetypes). If anything, the archetypes show that the process dimension of collective decision making features autonomous dynamics that impact the ability of actors to obtain and maintain fitness.

7.7 THE EVOLUTIONARY NATURE OF COLLECTIVE DECISION MAKING

Remember we started the chapter with presenting two opposing worldviews? One was founded on an entirely mechanistic understanding of decision making, that is, Clockworld, and the other on the premise of everything being fluid and random, that is, Cloudworld. While both worldviews have some utility in the science of collective decision making, they are also unhelpful because they either rely on gross simplification (Clockworld) or indulge in mystifying the ordinary (Cloudworld). We have made a modest attempt to develop and use a modified version of the fitness landscape as the proverbial Swiss army knife to enhance our understanding of collective decision making that combines the advantages of Clockworld (traceability) with those of Cloudworld (appreciation of specific conditions under which events take place). Our findings demonstrate that there are more patterns to be found than assumed in Cloudworld but also that each finding comes with a provision, which keeps us away from the kind of reasoning found in Clockworld. We solved this by presenting six archetypes and the rules of thumb associated with them. We are very much aware that there are probably other findings in the data we didn't address in this chapter, and we know that our approach can't explain *all* details of collective decision making. For example, we have no understanding of the personal motivations of individuals to act in certain ways. Here we run into the wall described by for example Dennett (1991), where adding more layers of data will lead to more elaborate descriptions but comes at the price of less discernible patterns.

What, then, is the evolutionary nature of collective decision making? Collective decision making is an uninterrupted and non-directional process that we structured in lineages of events. While non-directional, the process is limited to a certain (fluctuating) bandwidth that is determined by the configuration of substantive as well as relational aspects. Actors in the process try to gain fitness by seeking alliances with, or opposition to, other actors. Therefore the locus of fitness gains and losses is to be found in the interaction between those actors, as each action is carried out in response to actions of others. The consequences of these interactions are manifest in two principal forms of emergence. First, the aggregated result of all those individual actions in the field concerns changes in connections and problem and solution definitions, ultimately leading to changes in the fitness of each individual actor in the field. Fitness is context-dependent, which implies that it can be achieved through various strategies in response to what other actors do but also that, once obtained, it will be temporal. By and large, actors that have a high PSD and *c_score* are likely to achieve at least some of their goals, and actors that have a lower PSD and *c_score* can also be successful provided they are flexible. Inflexibility in combination with a low PSD and *c_score* is usually, but not by definition, not rewarded with fitness gains. Second, the interactions of individual actors combine to produce self-propagating dynamics that, once set in motion, drive the evolution of the collective decision-making process. We identified three principal forms here: force to fit, self-organized entrapment, and diversity breeds diversity. These dynamics are related to individual actions in a non-linear fashion, because those actions of individual actors contribute to, but cannot cause, the said dynamics. Taken together, these mechanisms characterize the evolutionary nature of collective decision making. The fitness field model presented in this book enables us to investigate the various dimensions of this process – ranging from individual strategies and actions to variation, selection and retention of contents, from interactions to fitness gains and losses, and back again.

We hope that our findings are firm enough to convince the reader that collective decision making is indeed evolutionary and, consequently, that evolutionary theories have tremendous utility for the analysis of such processes. This approach has the potential to negate the drawbacks of Clockworld and Cloudworld, a route we tried to clear in our research. The theoretical foundations are there already, and there is no direct need for conceptual innovation. However, researchers will be required to structure the messy reality of decision making without getting lost in the details, and to combine it with theoretical models that, while simplified by definition, are still in correspondence with that messy reality. Our fitness field model is developed with the express aim of sitting at that sweet spot. The final words are therefore for those actors in the real world trying to understand the situation they are in. To the actor feeling lost and struggling for fitness, we can recommend increasing connections as well as variety of problem and solution definitions. We can warn the winner that high fitness is temporal and can even backfire. We can tell the pragmatic actor to expect quick wins but also quick losses. We can give the conservative actor the advice that prudence may very likely not result in high fitness. We can advise the powerful actor that brute force can only lead to a selection of goals to be reached, that is, in the best case. Conversely, the powerless can derive solace from the fact that a good strategy in conjunction with favourable circumstances can return higher fitness. Evolution, then, is the most fundamental and important property of collective decision making.

Appendix A   Data processing and www.un-code.org

We developed a new application to store, structure, analyse and visualize the empirical data using the method and model described in this book. The application is called un-code.org, which stands for UNderstanding COllective DEcisions. Readers can access and use the application for free from the website with the same name. We will describe the details and use of the application in this appendix. The need for a new application rose from our requirements (which, in turn, can be traced back to the model and so on). The software we were looking for should: (1) not contravene the five prerequisites explained in Chapter 2; (2) provide a database function to store and retrieve raw qualitative data in a structured way; (3) allow the researcher to code and recode raw data for analytical purposes; (4) visualize the coded data in a three-dimensional fitness landscape; and (5) highlight specific dynamics in the landscape such as the adaptive walk. The application was developed by Julian Stieg.

A.1   WWW.UN-CODE.ORG APPLICATION DESCRIPTION

A.1.1   Software Architecture

www.un-code.org is open-source web-based software. It can be accessed online from anywhere in the world. To run it locally, the installation of a standard web service (such as the open-source program XAMPP) is needed beforehand. The software's back end is written in PHP5 and uses MySQL as the database system for storing user and case data in two separate databases. The front end of HTML5, JQuery and JavaScript utilizes the AJAX method – asynchronous server interactions – to achieve a fluid user experience. 3D visualization from within the browser is done using the Three.JS library operating the WebGL interface, thus enabling the software to use real 3D graphic acceleration from within any modern browser, such as Google Chrome, Mozilla Firefox, Apple Safari, Microsoft Edge or Internet Explorer 11. This approach allows dynamical and on-the-fly data visualization in a scalable environment. Although a dedicated video card is recommended for improved performance, graphics chips supporting OpenGL 2.1 or higher can also work.

The www.un-code.org software also provides the possibility of storing case study sources, such as documents or images up to the size of 32 Mbytes per file in any format, online on the server. The files are uploaded via the browser and stored in the MySQL database, which is only accessible from within a personal account, offering always-online availability and data protection at the same time.

A.1.2 Data Input

Before using the software, the user has to create a password-protected (stored encrypted in the database) user account. All entered data is *only accessible to the user*. In this section, we outline the order of steps needed for inputting data. The overview assumes that the empirical data for each component is readily available.

- Fields: The researcher defines the time span for each field, ranging from days to years per field. The time span cannot overlap with another field of the same lineage.
- Actors: The researcher then can assign actors by entering information for every field. A checkbox can be used to indicate if an existing actor is present or not in the subsequent fields. If it is present, it is included in the calculation and visualization. If the actor is (temporarily) absent, it will be taken out of that particular field.
- PSDs: The researcher needs to define the pool of problem and solution definitions for each lineage. Specific definitions can be assigned to each actor for every field, as of course definitions change throughout time. New definitions can be added later on, but the researcher needs to make sure that the definitions are assigned to the correct actors in the correct fields. This data is then used to compute the *PSD score* and is eventually utilized in the 3D visualization and persistence visualization features. The software can also generate a complete definitions overview as an Excel sheet.
- Connectedness: We differentiate between 'starting connections' forming the *starting c_score* and the qualified *weighted c_score*. To set the starting connections, the application creates a table of active actors for each field. One can then define a starting connection between two or more actors in a field by simply clicking checkboxes within that table. It is somewhat crude but much better than starting with randomization as is most often the case. The more refined *final*

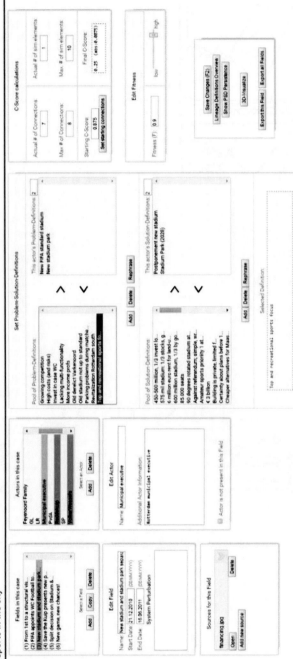

Notes: From left to right: fields/sources, actors, problem and solution definitions, *c_score* calculation/fitness. AJAX is used for every input.

Figure A.1 www.un-code.org's main data input mask

c_score is created by taking into account the similarities in elements of *PSDs*.

- Fitness: You can enter the fitness value by either typing the numerical value or using the slider. The fitness score per actor is represented on the *z*-axis of the 3D visualization. The score derives from the interpretation of the empirical data.

A.1.3   Calculations

www.un-code.org will automatically calculate values for *PSD* and *c_score*. These scores are calculated for every actor in every field. Any changes made to the case data update the calculation. This facilitates case study work, where such changes are often made during the research process. All score calculations will lead to results between 0 and 1 in order to facilitate comparability and scalability.

- PSD score

The *PSD score* for each actor per field is computed by the application as follows:

$$\text{every actor } i \text{ has } PSD_i = \frac{actual\ \#\ elements\ PSD_i}{maximum\ \#\ elements\ PSD}, \text{ where } PSD_i \in [0, 1]$$

Actual # elements PSD is number of elements in a PSD of an actor in the field. *Maximum # elements PSD* is the sum of all elements in all PSDs in the field. Inactive elements in PSDs that are present in the pool but not used are ignored by the application.

- c_score

The *starting c_score* is calculated as follows:

$$c\_score_i(t_0) = \frac{actual\ connections_i}{maximum\ \#\ connections}, \text{ where } c\_score_i \in [0, 1]$$

Actual connections are the number of connections of the actor to the other actors in this field. *Maximum # connections* concerns the maximum number of connections an actor can have, that is, connection with all active actors in this field. The weighted factor (*w*) is added as follows to the *final c_score*:

$$c\_score_i(t) = w_i \times \frac{actual\ connections_i}{maximum\ \#\ connections},$$

$$\text{where } w_i = \frac{\#\ actual\ similar\ elements\ PSD_i}{maximum\ \#\ similar\ elements\ PSD}, \text{ and } w_i \wedge c\_score_i \in [0, 1]$$

 # Actual similar elements PSD is the number of elements in the PSD of the actor that are shared with at least one other active actor in this field. *Maximum # similar elements PSD* is the maximum number of elements in the PSD that could be shared with at least one other active actor in this field.

A.1.4 Output Methods

Besides being a qualitative data management and calculation tool, www. un-code.org can visualize the data. There are two main outputs in the application: 3D visualization and persistence mapping.

3D visualization
The application offers the possibility of real-time 3D visualization within the browser. The data is rendered on three axes: PSD on the *x*-axis, CON (i.e. final *c_score*) on the *y*-axis and FIT on the *z*-axis. The application presents the results of PSD and CON in a relative way, with the highest value of the lineage being the ceiling of 1. For the PSD score, the researcher can select whether the ceiling should be relative to the whole lineage or only to the chosen field(s).

The researcher can then select which combination of fields and actors needs to be visualized. This allows the observation of a specific configuration at a specific point in time or, if multiple fields are selected, it allows monitoring the movement of one or multiple actors throughout the lineage. The movement can also be emphasized by the software using arrows to connect actors' positions in time. Different time positions are labelled with $t_1 \ldots t_n$. Earlier positions can be rendered semi-transparent for reading purposes.

The visualization can be rotated, tilted and zoomed or enlarged to allow for detailed examinations. The free 3D rotation can be done by holding the left mouse button; zooming is controlled by the mouse wheel. 3D labels for actors and grid will rotate so that they are always facing the viewer. The tool has a function to return to a standard original camera position in order to allow the researcher to compare various fields next to each other. It also has the function to take screenshots. Options in the visualization are: (1) coloured or greyscale output; (2) highlighting one or more specific actors in the field; and (3) sloped, peaked and columned fitness representation. Whenever selecting any option, changes to the visualization are applied instantly and no reload of the web page is required owing to the utilization of AJAX and WebGL abilities.

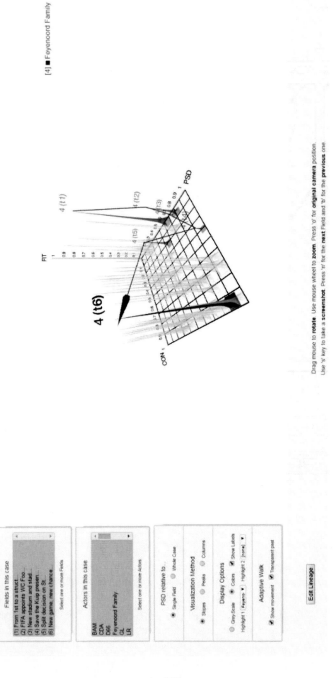

Figure A.2   www.un-code.org's 3D visualization: a typical case examination, highlighting the movement of an actor throughout multiple fields in relation to other actors

Persistence mapping

Over time, the interaction and alignment between actors in search of fitness produce a number of substantive evolutionary outcomes. Tracing the evolution of PSDs over time, and the positions and actions of actors tied to those PSDs, will give a thorough understanding of the mechanisms of variation, selection and retention at work. For this reason, www. un-code.org offers the possibility of mapping the persistence of PSDs throughout a lineage. The result per lineage is output as a *.xlsx sheet (using the library PHPExcel).

The fields of the selected lineage form the *x*-axis, and the definitions used anywhere in the lineage are presented on the *y*-axis, group-separated by 'problems' and 'solutions'. Each cell shows the percentage of actors that share this definition within this field. Next to the percentage, the exact number of actors sharing it and the total number of actors active in this field are shown. The strength of shared definitions is highlighted by different colours, with green representing non-shared or weakly shared definitions and red representing broadly shared definitions.

The columns on the right side show the number of total shares of this definition throughout the lineage, and the persistence score. The latter expresses in how many fields of the lineage a definition was active, that is, shared by at least one actor. This presentation allows the researcher to keep track of definitions getting weaker or stronger at specific points in time and helps point out possible explanation approaches as to why some definitions ended up being successful.

A.1.5 Coding and Interpretation

The one thing the application can't do is collect and code data. This has to be done by the researcher. It is also up to the researcher to assign certain values to certain data, that is, the coding of data. This implies that data has to be interpreted, which, in turn, means that validity and reliability are at stake. We took the following measures when processing the data presented in this book: (1) both authors coded all data and compared differences and similarities, which often led to changes; (2) data sources were triangulated in order to identify suspect sources and to inform the coding process; (3) data sources and coding have been made public in Appendix B and Appendix C. This enables the reader to check our interpretation. As always when working with qualitative data, one can disagree about an interpretation, but such disagreement should not concern the procedure.

un-code.org PSD persistence of lineage "Sports in the city"	From 1st to a structural vision	FIFA appoints WC Football to Russia	New stadium and stadium park separated	Save the Kulp presents new plan	Split decision on Stadium & Stadium Park	New game, new chances!	Total Shared	Persistence
Problem Definitions								
Growing competition	11% (1/9)						1	1
High costs (and risks)			89% (8/9)	36% (4/11)	73% (8/11)		20	3
Invest in case WC							0	0
Lacking multi functionality	11% (1/9)						1	1
More income profs						20% (1/5)	1	1
New FIFA standard stadium		89% (8/9)	100% (9/9)	100% (11/11)	100% (11/11)	100% (5/5)	44	5
New stadium park		11% (1/9)	11% (1/9)	9% (1/11)			3	3
Old derelict Varkenoord				9% (1/11)			1	1
Old stadium not up to standard	100% (9/9)						9	1
Parking problems during matches		11% (1/9)					1	1
Revitalization Rotterdam south	89% (8/9)						8	1
Top and recreational sports focus	89% (8/9)						8	1
Solution Definitions								
450-500 million, 1/3 invest loc gov		11% (1/9)		9% (1/11)			1	1
575 mil stadium: 1/3 stocks, gov							0	0
6 million euro rent for land-use		11% (1/9)					1	1
600 million stadium, 1/3 by government	11% (1/9)	22% (2/9)					2	1
85,000 seats							2	1
90 degrees rotated stadium at Maas		78% (7/9)		9% (1/11)			7	1
Against referendum; simple, wrong quest							3	2
Amateur sports priority 1 at Varkenoord				9% (1/11)	18% (2/11)		3	2
€ 3 billion	11% (1/9)						1	1
Building is private, limited funding gov	67% (6/9)						6	1
Certainty about plans before 1-1-2014				9% (1/11)			1	1
Cheaper alternatives for Maas variant			44% (4/9)	9% (1/11)	9% (1/11)		6	3
Financing of metro line by city region	44% (4/9)						4	1
Gov. finance under very strict condition				18% (2/11)			2	1
Independent fin, risk assessment stadium		11% (1/9)					1	1
Independent research risk, fin perspecti		78% (7/9)					7	1
Invest in case WC, nat gov coinvest		11% (1/9)					1	1

Figure A.3   www.un-code.org's persistence mapping, showing which definitions survived over time

Appendix B   Data collection

This appendix provides an overview of the data sources used for the empirical studies presented in Chapters 5 and 6. The full list of references includes about 1800 sources and can't be published here because of practical limitations. This list is available from the authors upon request.

B.1   DATA COLLECTION FOR HSL-ZUID

For the HSL-Zuid study, we collected both newspaper articles and policy documents to trace the lineages of the study, the movements of the actors and their considerations for undertaking certain actions. Dutch newspaper articles published by *Algemeen Dagblad*, *Financieele Dagblad*, *NRC Handelsblad*, *Reformatorisch Dagblad*, *Telegraaf*, *Trouw* and *Volkskrant* were retrieved from the LexisNexis database. We started the search with five different generic themes that covered the whole decision-making process. This resulted in the following overview of articles:

Generic search	Specific strand	Period	No. of articles
HSL-Zuid	Building, construction, etc.	28 April 1994 – 23 December 2009	321
	Tender, infrastructure, etc.	17 February 1995 – 2 July 2004	183
	Concession, operation, etc.	10 July 2001 – 4 April 2007	67
	Fyra, high-speed train, etc.	12 July 2005 – 22 November 2008	38
	Operation, V250, Fyra, etc.	8 September 2009 – 5 March 2013	577
			1186

In addition to the newspaper articles, we retrieved six in-depth articles about the HSL and/or V250 from (technical) railway journals not covered in the LexisNexis database. Some sources appeared multiple times because of the predefined structure in five strands. In addition, we triangulated the articles from multiple sources, which led to the rejection of a small number of articles when we thought that they were unreliable or contradictory. The selection is skewed towards more recent years, as news coverage was more intensive then than during the early stages of the project. It also appeared that recent newspaper articles showed more similarities than older ones. Redundant articles have been removed, which has corrected the skewed distribution somewhat.

The articles were then structured in an event-sequence database (Spekkink, 2015) and then coded to reconstruct how different events evolved throughout the time of the study. This made it possible to select the significant events to form the lineages presented in Chapter 5. The event sequence is a condensed presentation of the news articles, as certain events were covered by several newspapers; in total 412 events were identified as significant for the collective decision-making process. The exact dynamics of those events were fleshed out on the basis of government papers: 6 policy documents of the Ministry of Infrastructure, 27 letters of Parliament and 28 evaluation reports (from the Court of Audit, temporary evaluation committees, etc.). This enabled us to reconstruct the decisions actors took, the context of those decisions and the rationale for the decisions.

The writing of the book coincided with a parliamentary inquiry regarding the Fyra debacle, to which we also made a minor contribution. The report was published in 2015 as 'Reiziger in de kou' (Tweede Kamer der Staten-Generaal, kst-33678-10). The publication led to considerable media coverage. We have taken notice of the news coverage and the report, and used the information to cross-check our findings. It confirmed our data analysis, and therefore no amendments to our study were needed.

B.2   DATA COLLECTION FOR GOTTHARD

Data for this case was also collected from newspaper sources, most prominently the *Neue Zürcher Zeitung* and swissinfo.ch, and from policy papers. We searched for 'future Gotthard region', 'consequences GBT' and related terms. The media coverage of the cooperation progress leading up to PREGO/San Gottardo was considerably less extensive than in the other studies, in particular in comparison to the news coverage of the Gotthard Base Tunnel itself. In total, we used 23 newspaper articles. Given the nature of the issue and the lack of coverage, we needed to rely on government papers: 18 policy and evaluation reports, including reports from the Bundesrat and the cantons, as well as project reports. Since a search in LexisNexis didn't return the desired results, we relied primarily on Google to retrieve the documents. We were assisted by a German speaker, Julian Stieg, in searching for and translating the data. The structuring of the data was checked again by Julian Stieg, which led to some final changes.

B.3   DATA COLLECTION FOR SPORTS IN THE CITY

Data was collected from Dutch newspaper articles, mostly published by *Algemeen Dagblad/Rotterdams Dagblad*, retrieved from LexisNexis. The search terms used were 'nieuwe kuip', 'Feyenoord stadium', 'sportpark' and 'stadionpark'. The search resulted in 412 articles that cover the full duration of the study. The articles were structured using the event-sequence database mentioned above and then coded to reconstruct the lineage. The final lineage featured 105 events. In addition to the written sources, we carried out 10 interviews with key people involved in Stadium Park and/or the stadium. The interviews were conducted by Lasse Gerrits and Iris Korthagen (except the one with Buijsen and Pakasi, which was done by Gerrits and Danny Schipper), and are as follows:

Respondent	Function	Date
Mark van den Boer	Spokesperson, municipality of Rotterdam	23 February 2011
Hans van Rossum	Project manager, Stadium Park	15 November 2012
Marieke Gruijthuisen	Spokesperson for the Alderman of Sports	15 November 2012
Jan van Merwijk	Director, Feyenoord Stadium	22 November 2012
Robin van Holst	Save De Kuip group	18 December 2012
Kirsten Verdel	Ice Skating Club Rotterdam	20 February 2013
Gert Onnink	Sports journalist, *Algemeen Dagblad*	21 February 2013
Jan Geuskens	Former project manager, Stadium Park	21 February 2013
Ard Buijsen and Erwin Pakasi	Voluntary advisers, Feyenoord family	1 March 2013
Aad van der Laan	Former president, Feyenoord Sports Club	10 April 2013

We also used three policy documents published by the municipality of Rotterdam (Structural Vision, Masterplan Sports Campus, and Environmental Impact Assessment), several reports (i.e. evaluation reports, master's theses), and internet sites. The reconstruction method was the same as for the studies listed above.

B.4   DATA COLLECTION FOR BANGKOK

Most of the written data came from English newspaper articles published in the *Bangkok Post*, the *Nation* and the *Railway Gazette*, and was retrieved from LexisNexis. The search terms used were 'Bangkok', 'Makkasan', 'Suvarnabhumi', 'Lat Krabang', 'Hopewell' and 'Airport Railway Link'. We were limited to English sources because we couldn't engage a Thai translator, though this may not have had much impact on the results, because the *Bangkok Post* and the *Nation* are both reputable newspaper publishers from Bangkok. The search resulted in 114 useful articles. In addition to the written sources, we carried out six interviews during a site visit that took place between January and March 2014. All interviews were conducted in English by Lasse Gerrits and are as follows:

Respondent	Function
Dr Sumet Ongkittikul	Researcher at Thailand Development and Research Institute
Dr Waressara Weerawat	Researcher at Mahidol University, Logistics Innovation Centre
Dr Jirapan Liangrokapart	Researcher at Mahidol University, Logistics Innovation Centre
Dr Krit Anurakamonkul	Superintending engineer, State Railway of Thailand (two meetings)
Dr Chumloon Tangpaisalkit	Chairman, Board of Directors, State Railway of Thailand Electrified Train

In addition, several research papers (i.e. Property Rights Bangkok, Built Environment Behaviour, Railway Sector Reform Studies) and internet sites were used to reconstruct the contextualized considerations of actors and their subsequent actions.

Appendix C   Data-coding the high-speed railway study

We present the interpretation and coding of the data for the high-speed railway study here in full so that the reader can trace how we moved from raw data to analysis and conclusions. The overview is structured in the four lineages.

C.1   LINEAGE 1 – THE FINANCES OF BUILDING THE HSL-ZUID

Field 1: Building Starts and Fraud Detected

From the start, all the builders had a joint working history. Additionally, the fraud case revealed close connections on all fronts. Thus they were all connected. The Minister of Transport had working relations with all the actors, while the House of Representatives had a history only with the Minister, as is shown here:

Actor *c_score*	Ballast	Bechtel	B/K	MT	HoR	NBM	VW S
Ballast Nedam		x	x	x		x	x
Bechtel	x		x	x		x	x
Bouygues/Koop	x	x		x		x	x
Minister of Transport	x	x	x		x	x	x
House of Representatives				x			
NBM Amstelland	x	x	x	x			x
VolkerWessels Stevin	x	x	x	x		x	

Table C.1   Problem and solution definitions in lineage 1, field 1

Actor PSD		Problem definition		Solution definition
Ballast Nedam	1.	Building rails, etc.	1.	Schiphol–Green Heart Tunnel and Rotterdam–Moerdijk (February 2000)
Bouygues/Koop	1.	Building tunnel	1.	Tunnel; single tube (diameter 15 metres) (February 2000)
Holzmann	1.	Building rails, etc.	1.	Brabant South (February 2000)
House of Representatives	1.	Finances HSL	1.	Astonished about agreements with builders, and rising finances (14 October 2000)
			2.	Investigation into finances and builders' agreement (14 October 2000)
Minister of Transport	1.	Building HSL-Zuid	1.	Tender of foundation in six separate parts (design and construct) (6 May 1999)
	2.	Finances HSL-Zuid	2.	Tender rails, catenary and signal system through DBFMO (6 May 1999)
			3.	Contract with five consortia: €1.9 billion (14 March 2000)
			4.	Tendering for catenary, etc. (two consortia still in the running) (20 October 2000) → half-year later Infraspeed wins (5 May 2001)
			5.	Longer negotiations with builders, acquiring land (22 July 2001)
NBM Amstelland	1.	Building rails, etc.	1.	Green Heart–Rotterdam (February 2000)
VolkerWessels Stevin	1.	Building rails, etc.	1.	Brabant North (February 2000)

Solution: The Dutch Authority for Consumers and Markets announced fines for the building companies that had made price agreements in the preceding years, with the fines ranging from €2000 to €19 million (5 April 2005). However, it was not made public which firms in the HSL tender were fined what amount. This created a great deal of change in the HSL-Zuid case; questions regarding who had to pay for what, and who was to blame started to play a role.

C.2 LINEAGE 2 – ROUTE AND TRACK DECISIONS FOR HSL-ZUID

Field 1: Crossing Borders and a Tunnel

As all the actors involved in this field were political and/or administrative bodies, they had a history of links connecting them. The exception here was the Belgian Minister, who had a working history only with the Dutch Minister. This created the following starting connections as input for the *c_score*:

Actor *c_score*	BMoT	MT	HoR
Belgian Minister of Transport		x	
Minister of Transport	x		x
House of Representatives		x	

Table C.2   Problem and solution definitions in lineage 2, field 1

Actor PSD	Problem definition	Solution definition
Belgian Minister of Transport	1. Route across border	1. Route via Roosendaal (7 October 1995) 2. Route via Breda (E19) including compensation for extra cost (24 May 1996)

Minister of Transport	1. Route HSL-Zuid	1. Route via Breda (7 October 1995)
		   a. Compensation for Belgium (24 May 1996)
		2. Tunnel through Green Heart (April 1996)
		   a. Compensation for villages (€400 million)

House of Representatives	1. Route HSL-Zuid	1. Green Heart Tunnel (April 1996) a. Compensation for villages

Field 2: Versions of the Green Heart Tunnel

The Minister was connected to all the actors involved in this field, as they were political and/or administrative bodies and had a history of links. This created the following starting connections as input for the *c_score*:

Actor c_score	HoR	MT	Villages
House of Representatives		x	
Minister of Transport	x		x
Villages (stakeholders)		x	

Table C.3   Problem and solution definitions in lineage 2, field 2

Actor PSD	Problem definition	Solution definition

Minister of Transport	1. Building HSL-Zuid
	2. Tunnel length
		1. Tunnel 6 or 9 km (October 1996)
		2. Against dug tunnel, needs adjustment planning decision (22 July 1998)
		3. Tender (DBFMO) for drilled tunnel of 6.4 or 2.6 km (February 1999)
		  a. Safety and comfort of passengers
		  b. 300 km/h
		  c. No pressure waves
		4. Longer tunnel option, as short is harmful (20 April 1999)

CDA and VVD	1. Tunnel length	1. Tunnel dug instead of drilled (22 July 1998)
Villages (stakeholders)	1. Tunnel length	1. Long tunnel, i.e. 6.4 km (February 1999)

Solution: On 22 December 1999 the HSL project bureau presented plans for the new Green Heart Tunnel: a single tube with a diameter of almost 15 metres, except in the middle of the tunnel, which would hold a small section with two tubes. This diameter was necessary to avoid the effects of the pressure wave on passengers. The French/Dutch consortium Bouygues/Koop would build the tunnel for €430 million (connection with lineage 1).

Field 3: Building and Safety

The only two actors in this field were the House of Representatives and the Minister of Transport, which were by definition connected, as they had a shared history:

Actor c_score	HoR	MT
House of Representatives		x
Minister of Transport	x	

Table C.4   Problem and solution definitions in lineage 2, field 3

Actor PSD	Problem definition	Solution definition
House of Representatives	1. Building HSL	1. No pergola, as it is contradicts nature preservation (13 July 2000) 2. Stop building due to heavy metals (20 December 2001)
Minister of Transport	1. Building HSL	1. Pergola a. Willing to abandon pergola if province and municipalities contribute (12 September 2000); then lower crossing is possible (3 October 2000) 2. Lowered track Bergschenhoek (€20 million more) and builders receive nine months' extra building time (22 June 2001) 3. Remove furnace slacks and place new ones (21 February 2002) 4. ERTMS is the only safety system (24 May 2004), but is delayed for testing (3 May 2006) and upgraded (20 December 2006)

Solution: Except for minor incidents, for example dust accumulation in the tunnel, modifications of noise barriers, and so on, the construction continued, and on 14 October 2008 the HSL-Zuid was ready for operation. The train should be able to travel from Amsterdam to Brussels (212 kilometres) in 1 hour and 46 minutes with an average speed of 120 km/h including stops.

C.3   LINEAGE 3 – CONCESSION FOR OPERATING THE HSL-ZUID

Field 1: From Favoured Party to One of Many

The House of Representatives and the Minister of Transport were by definition connected, as they had a shared political history. The Minister was also connected to NS, because NS used to be a state agency and also because they had worked together on many occasions.

Actor c_score	HoR	MT	NS
House of Representatives		x	
Minister of Transport	x		x
NS		x	

Table C.5   Problem and solution definitions in lineage 3, field 1

Actor PSD	Problem definition	Solution definition
House of Representatives	1. Concession HSL	1. Minister is too pro-active in going for exclusive rights for NS and should keep options open (7 April 1998) 2. Different opinions on NS as sole provider and given deadline for new proposal (13 November 1999)

Minister of Transport	1. Concession HSL	1. Cutting up rail net into regional and central net
		   a. HSL part of central net, hence open to parties other than NS as well (5 February 1998)
		 2. Exclusive rights for HSL to NS (23 March 1999)
		   a. Registering NS for stock market, regional nets for tender if NS doesn't want them
		 3. Start tender and NS should withdraw offer (19 June 1999)
		   a. Tender will be in favour of NS, owing to reciprocity
		 4. Offer by NS (1 September 1999) inadmissible (13 November 1999)
		   a. Joint Dutch and foreign bid
		   b. Offer €360 million too low
		   c. NS will get one week for renewed offer

NS	1. Concession HSL	1. Sole right to operate (22 June 1998) 2. Applying to stock market for finances 3. Makes offer for exclusive right (1 September 1999) 4. Not willing to make new offer following dictate of Minister (17 November 1999)

Solution: On 17 November 1999 NS announced that it was not willing to follow the dictate of the Minister and would not submit a new offer. The same day the Minister announced that the operation of HSL would now be put out to tender, because NS had refused to provide a better offer.

Field 2: Disinformation, but Still Tendering

The connections were the same as in the previous field:

Actor *c_score*	HoR	MT	NS
House of Representatives		x	
Minister of Transport	x		x
NS		x	

Table C.6   Problem and solution definitions in lineage 3, field 2

Actor PSD	Problem definition	Solution definition
House of Representatives	1. Concession HSL	1. No tender because Minister misinformed House (9 December 1999)

Minister of Transport	1. Concession HSL	1. Will delay tender (15 December 1999)
		   a. Maybe NS state-owned again
		   b. NS maybe eligible monopoly
		   c. In event of tender, NS will win
		2. Bid by NS consortium is admissible, but decision is that exclusive offer cannot be made (EU regulation) (4 May 2000)

NS	1. Concession HSL	1. NS, together with KLM and Schiphol, makes offer (4 April 2000)

Solution: On 4 May 2000 the Cabinet decided that NS, KLM and Schiphol could not make an exclusive offer and put the concession for HSL-Zuid up to tender, this time a completely open one.

Field 3: NS Wins Public Tender for HSL-Zuid

This was a fairly simple field. The Minister of Transport was connected to all the actors, as the Minister was the actor putting the concession out to tender. All the other actors knew each other but were competitors in the tendering process. As such they were not connected in this field.

Actor *c_score*	HoR	MT	NS	Conn	DB	SC
House of Representatives		x				
Minister of Transport	x		x	x	x	x
NS, KLM and National Express		x				
Connexxion, CGEA and SJ		x				
DB and Arriva NL		x				
Stagecoach		x				

Table C.7   Problem and solution definitions in lineage 3, field 3

Actor PSD	Problem definition	Solution definition
Connexxion, CGEA and SJ	1. Concession HSL	1. €61 million (June 2001)
DB and Arriva NL	1. Concession HSL	1. €100 million per year is too high (3 February 2001) 2. The offer for the bid is €100 million (June 2001)
House of Representatives	1. Concession HSL	1. No tender, because of fear of big foreign contenders (5 May 2000) 2. Allows tender, but still fears big foreign contenders (29 June 2000)

Minister of Transport	1. Concession HSL	1. Public tender, with requirements such that NS will win (14 June 2000):
		    a. Dutch office, and knowledge of Netherlands rail
		    b. Level playing field (reciprocity)
		2. Independent committee to evaluate bids (2 October 2000)
		3. Four bids qualify and receive bid information (26 October 2000)
		4. At least €100 million for concession for 15 years (in the event of no proper offer, state will operate itself) (25 January 2001)
		5. NS will receive concession (16 June 2001); Connexxion is second, and DB and Arriva third

NS, KLM and National Express	1. Concession HSL	1. €148 million (June 2001)
Stagecoach	1. Concession HSL	1. Withdraws, because it does not have a Dutch partner (3 May 2001)

Solution: On 10 July an agreement on the main issues for the concession of HSL-Zuid was reached between the Minister and NS. The concession cost was €148 million per year (established after negotiation); the contract would be valid for 15 years, and required 96 services per day.

Field 4: Price Disputes

The Minister of Transport was connected with NS (HSA) and the House of Representatives, as in the previous fields:

Actor c_score	HoR	MT	NS
House of Representatives		X	
Minister of Transport	X		X
NS (HSA)		X	

Table C.8   Problem and solution definitions in lineage 3, field 4

Actor PSD	Problem definition	Solution definition
House of Representatives	1. Concession HSL	1. Against high price rise tickets (5 December 2001) a. Renegotiate contract with HSA for lower ticket price (29 August 2002)
Minister of Transport	1. Concession HSL	1. Prices 50 per cent higher for HSL-Zuid (5 December 2001) 2. No maximum on prices for tickets (29 August 2002), reconfirmed 4 October 2002) based on information from HSA

HSA	1. Concession HSL	1. Raise ticket price 50 per cent to cover cost of concession (5 December 2001)
		2. Two bid options (29 June 2002):
		  a. €148 million, high ticket price (60 per cent higher than normal)
		  b. €101 million, lower ticket prices (25 per cent higher than normal)

Solution: On 4 October the Minister reconfirmed that he would not request a maximum price for tickets. This decision was based on the information provided by NS (HSA) that a lower price would increase the number of services on the route. In other words, the concession stayed at €148 million with no price restriction. On 12 December the House of Representatives wanted the Minister to request more information from NS (HSA) to assess if it was really impossible to have a maximum price for the tickets. The Minister promised to look into it.

Field 5: Lost Time and Lost Connections between the Netherlands and Belgium

The Minister of Transport was connected with NS (HSA) as in the previous fields. The Minister was also connected with the Belgian Minister of Transport as a result of having had many related issues and a working history. The Belgian Minister was connected with the Belgian train operator NMBS, as they had been working together for years. NS and NMBS were connected, as they jointly operated HSL-Zuid, but also as they had been servicing cities across the border together.

Actor *c_score*	Be MT	NS	MT	NMBS
Belgian Minister of Transport			x	x
NS (HSA)			x	x
Minister of Transport	x	x		
NMBS	x	x		

Table C.9　Problem and solution definitions in lineage 3, field 5

Actor PSD	Problem definition	Solution definition
Belgian Minister of Transport	1. Servicing cities 2. Deepening Westerschelde	1. Recover lost time, service Breda and Den Haag, and prioritize HSL on the tracks (1 June 2004)
Minister of Transport	1. Belgians not servicing cities and recovering delay	1. No renegotiation concession with NS, as Belgium is responsible for lost time and no service to Breda and Den Haag (19 March 2004) 　　a. HSA and NMBS need to find joint solution 2. Different schedule to prioritize HSL (1 June 2004) 3. No deepening of Westerschelde as long as Belgian Minister does not put pressure on NMBS (21 December 2004)
NS	1. Servicing cities	1. Reduction on concession cost as Belgium has made wrong time calculations (17 minutes) (19 March 2004)
NMBS	1. Servicing cities	1. No service, as risk of losses (19 March 2004)

Solution: On 12 March 2005 the two ministers reached an agreement on HSL-Zuid: the Dutch Minister accepted a loss of eight minutes' travelling time in exchange for the promise of shuttle services to Breda and Den Haag.

Field 6: NS (HSA) Uses Monopoly Power, but Still Needs to Be Saved from Bankruptcy

The Minister of Transport was connected with NS (HSA) as in the previous fields:

Actor *c_score*	MT	NS
Minister of Transport		x
NS (HSA)	x	

Table C.10   Problem and solution definitions in lineage 3, field 6

Actor PSD	Problem definition	Solution definition
Minister of Transport	1. Concession HSL	1. Bankruptcy of NS versus losses of new concession (31 January 2011) 2. New concession realizes €2.2 billion (loss to Minister of Transport €390 million, as against €2.4 billion in the event of bankruptcy) (18 November 2011)
NS	1. Concession HSL	1. Lower concession fee (9 December 2010) a. Lasts for 15 years b. Cost of €130/km

Solution: On 18 November 2011 the new concession realized €2.2 billion (loss to Minister of Transport €390 million, as against €2.4 million in the event of bankruptcy).

Field 7: Finally Trains Going to Breda and Den Haag

The Minister of Transport was connected with NMBS as a result of decisions in previous fields:

Actor c_score	MT	NMBS
Minister of Transport		x
NMBS	x	

Table C.11   Problem and solution definitions in lineage 3, field 7

Actor PSD	Problem definition	Solution definition

Minister of Transport	1. Servicing cities	1. No new schedule, as NMBS should service Breda and Den Haag (19 September 2012)
		  a. NMBS should buy new train
		2. Possible legal steps against NMBS (10 October 2012)
		3. Financial contribution (€2.5–3.5 million) (4 December 2012)
		  a. High-speed service Breda to Antwerp
		  b. Intercity connection Den Haag to Rotterdam

NMBS	1. Servicing cities	1. No service to Breda and Den Haag (22 November 2011) a. Not profitable 2. Dutch government should financially assist (19 September 2012)

Solution: On 4 December 2012 it was agreed between the Minister of Transport, NS and NMBS that a high-speed train would run between Breda and Antwerp. Den Haag would be serviced by an Intercity to Rotterdam, where passengers could change to the Fyra services. The Dutch government would contribute financially to make this possible, at an estimated cost of between €2.5 million and €3.5 million. NS and NMBS were happy, but traveller organizations Rover (Netherlands) and TreinTramBus (Belgium) were dissatisfied, as it was a compromise between nations which they claimed wouldn't benefit passengers.

Field 8: Back to Square 1

The Minister of Transport was connected with NS (HSA) as in the previous fields:

Actor *c_score*	MT	NS
Minister of Transport		x
NS (HSA)	x	

Table C.12   Problem and solution definitions in lineage 3, field 8

Actor PSD	Problem definition	Solution definition

Minister of Transport	1. Servicing HSL
	2. Servicing Den Haag to Brussels
		1. Alternatives for HSL, since no V250 (22 January 2013)
		2. Penalties for NS and NMBS if they don't fulfil contract (22 January 2013)
		   a. Punctuality
		   b. Passenger satisfaction
		3. Temporary Intercity service Den Haag to Brussels will remain (26 February 2013)
		4. No new contracts, or new tender, for HSL-Zuid (26 February 2013)

NS (and NMBS)	1. Servicing HSL	1. New scenarios about servicing HSL-Zuid (26 February 2013)

Solution: Replacement Intercity trains would service the line.

C.4 LINEAGE 4 – OPERATING THE HSL-ZUID

Field 1: Tendering Rolling Stock

NS (HSA) was connected to all the actors, as it was the actor putting out the tender for new trains. All the train builders knew each other but were competitors as a result of the tendering process. As such they were not connected in this field. The Minister of Transport was connected to NS (HSA), as they were working together in several other fields, and had a related working history.

Actor c_score	Alstom	AB	Bom	NS (HSA)	MT	Siemens
Alstom				x		
AnsaldoBreda				x		
Bombardier				x		
NS (HSA) (and NMBS)	x	x	X		x	x
Minister of Transport				x		
Siemens				x		

Table C.13   Problem and solution definitions in lineage 4, field 1

Actor PSD	Problem definition	Solution definition
Alstom	1. Rolling stock	1. Double-decker AGV, price per seat €34 384 (autumn 2003) 2. Withdraw (23 December 2003)
AnsaldoBreda	1. Rolling stock	1. New train, 220 km/h and 546 seats 2. V250, price per seat €34 631 (autumn 2003)
Bombardier	1. Rolling stock	1. Locomotives 200 km/h, coaches variable 2. Withdraw (summer 2003)
NS (HSA) and NMBS	1. Rolling stock	1. 23 trains ready by October 2006 (19 April 2002) 　a. 220 km/h 　b. Price per seat should be low 　c. 450–550 seats 2. Only 12 trains (with option for 14 more), new offer requested from Alstom and Bombardier (23 December 2003)
Minister of Transport	1. Rolling stock	1. 250 km/h due to required service time (summer 2003)
Siemens	1. Rolling stock	1. Velaro, 300 km/h 2. Withdraw (summer 2003)

Solution: In autumn 2003 the V250 option offered by AnsaldoBreda was selected as the winner of the tender.

Field 2: Troubles with ERTMS

The House of Representatives was connected to the Minister, whereas the Minister was also connected to NS. NS was also connected to the train builder AnsaldoBreda:

Actor *c_score*	AB	HoR	NS (HSA)	MT
AnsaldoBreda			x	
House of Representatives				x
NS (HSA)	x			x
Minister of Transport		x	x	

Table C.14   Problem and solution definitions in lineage 4, field 2

Actor PSD	Problem definition	Solution definition

AnsaldoBreda	1. ERTMS
	  a. Constantly adjusted norms
	  b. Conflicting laws between the Netherlands, Belgium and the EU
		1. Train delivery to be delayed as a result of ERTMS issues (16 December 2005); trains will probably be more expensive

House of Representatives	1. Operating HSL	1. Does not keep NS to its contract, as it could go bankrupt (2 November 2005) 2. Demands solution from Minister about ERTMS and delayed delivery of trains, to be paid for by NS (11 November 2005) 3. NS should fix late delivery, number of expected passengers and ERTMS (based on report by Court of Audit, 20 July 2007)

Minister of Transport	1. Operating HSL
	2. Late delivery of trains
		1. Only ERTMS (12 October 2005)
		2. NS to be held to its contract (€148 million/year) (12 October 2005)
		3. NS should lease trains in the meantime (12 October 2005)
		  a. NS should pay for the leased trains
		4. No legal battle with and no deadline for NS, as the Minister has underestimated ERTMS (25 April 2007)
		5. SA will be compensated
		  a. Due to late delivery of ERTMS by Minister
		  b. Unclear starting date for HSL, also due to dust formation in Green Heart Tunnel (11 September 2007)

Table C.14   (continued)

Actor PSD	Problem definition	Solution definition
NS (HSA)	1. Operating HSL 2. Later delivery of trains due to ERTMS	1. Another safety system besides ERTMS (12 October 2005) 2. Leasing locomotives (160 km/h) till V250 is delivered, saving travelling time of 10 minutes instead of 30 (7 November 2005), i.e. 12 Bombardier Traxx locomotives (27 December 2005) 3. Charge Minister for leasing locomotives from Bombardier (27 December 2005)

Solution: In a letter to the House the Minister announced that the Minister was found responsible for the late delivery and would discuss compensation with NS (HSA) (11 September 2007). In this letter he also wrote that it was unclear whether the starting date of HSL-Zuid in December 2007 could be met. The first reason was the ERTMS problems, but there was also excessive dust accumulation in the Green Heart Tunnel. Besides these building issues he was concerned that AnsaldoBreda would not deliver the trains on time, but that the leased Traxx locomotives were not yet fitted with ERTMS.

References

Abbott, A. (2001), *Time Matters: On Theory and Method*, Chicago: University of Chicago Press.

Abell, P. (1984), Comparative narratives: Some rules for the study of action, *Journal for the Theory of Social Behaviour*, *14*(3), 309–331.

Abell, P. (2004), Narrative explanation: An alternative to variable-centered explanation?, *Annual Review of Sociology*, *30*(1), 287–310.

Alberch, P. (1994), Review of *The Origins of Order*, *Journal of Evolutionary Biology*, *7*, 518–519.

Alchian, A. (1950), Uncertainty, evolution, and economic theory, *Journal of Political Economy*, *58*(3), 211–221.

Algemene Rekenkamer in Tweede Kamer der Staten-Generaal (2003), *Rapport risicoreservering HSL-Zuid en Betuweroute*, Den Haag: SDU.

Altenberg, L. (1997a), Genome growth and the evolution of the genotype–phenotype map, in W. Banzhaf and F.H. Eckman (eds), *Evolution and Biocomputation: Computational Models of Evolution* (Lecture Notes in Computer Science 899, pp. 205–259), Berlin: Springer.

Altenberg, L. (1997b), *NK* fitness landscapes, in T. Bäck, D.B. Fogel and Z. Michalewicz (eds), *The Handbook of Evolutionary Computation* (pp. 1–11), Oxford: Oxford University Press.

Arrow, K.J. (1994), Methodological individualism and social knowledge, *American Economic Review*, *84*(2), 1–9.

Arthur, W.B. (1994), *Increasing Returns and Path Dependence in the Economy*, Ann Arbor: University of Michigan Press.

Arthur, W.B. and Durlauf, S.N. (1997), *The Economy as an Evolving Complex System II* (Santa Fe Institute Series), Santa Fe, NM: Santa Fe Institute.

Atran, S. and Norenzayan, A. (2004), Religion's evolutionary landscape: Counterintuition, commitment, compassion, communion, *Behavioral and Brain Sciences*, *27*(6), 713–730.

Auerswald, P., Kauffman, S., Lobo, J. and Shell, K. (2000), The production recipes approach to modeling technological innovation: An application to learning by doing, *Journal of Economic Dynamics and Control*, *24*(3), 389–450.

Axelrod, R. (1984), *The Evolution of Cooperation*, New York: Basic Books.

Axelrod, R. (1986), An evolutionary approach to norms, *American Political Science Review*, *80*(4), 1095–1111.

Axelrod, R. (1997), *The Complexity of Cooperation: Agent-Based Models of Competition and Collaboration*, Princeton, NJ: Princeton University Press.

Axelrod, R. and Bennett, D.S. (1993), A landscape theory of aggregation, *British Journal of Political Science*, *23*(2), 211–233.

Babcock, H.M. (1996), *Democracy's Discontent* in a complex world: Can avalanches, sandpiles, and finches optimize Michael Sandel's civic republican community?, *Georgetown Law Journal*, *85*, 2085–2103.

Bacaër, N. (2010), *A Short History of Mathematical Population Dynamics*, London: Springer.

Barr, J. and Hanaki, N. (2008), Organizations undertaking complex projects in uncertain environments, *Journal of Economic Interaction and Coordination*, *3*(2), 119–135.

Bartels, J. (1987), *Kennis, geschiedenis, objectiviteit: Een filosofische reflectie op enkele ontwikkelingen in de wetenschapstheorie* (1st edn), Groningen: Konstapel.

Barton, N.H. (2005), Fitness landscapes and the origin of species, *Evolution*, *59*(1), 246–248.

Bau-, Verkehrs- und Forstdepartement Graubünden (2005), 'Raumkonzept Gotthard: Grundlagen, Inhalte, Struktur und Prozess', Policy paper.

Baumgartner, F.R. and Jones, B.D. (1993), *Agendas and Instability in American Politics* (1st edn), Chicago: University of Chicago Press.

Becker, M.C., Knudsen, T. and Stieglitz, N. (2008), Surfing on the good waves: How firms benefit from changing market conditions, Paper presented at the 25th Celebration Conference 2008 on Entrepreneurship and Innovation – Organizations, Institutions, Systems and Regions, CBS, Copenhagen, 17–20 June.

Bednarz, J. (1984), Complexity and intersubjectivity: Towards the theory of Niklas Luhmann, *Human Studies*, *7*(1–4), 55–69.

Beinhocker, E. (1999), Robust adaptive strategies, *MIT Sloan Management Review*, *40*(3), 95–106.

Beinhocker, E. (2006), *The Origin of Wealth: Evolution, Complexity and the Radical Remaking of Economics*, Boston, MA: Harvard Business School Press.

Bhaskar, R. (1979), *The Possibilities of Naturalism*, Atlantic Highlands, NJ: Humanities Press.

Bhaskar, R. (2008), *A Realist Theory of Science*, New York: Routledge.

Bijker, W.E. (1997), *Of Bicycles, Bakelites, and Bulbs: Toward a Theory of Sociotechnical Change* (reprint edn), Cambridge, MA: MIT Press.

Bocanet, A. and Ponsiglione, C. (2012), Balancing exploration and exploitation in complex environments, *VINE*, *42*(1), 15–35.

Bolinska, A. (2013), Epistemic representation, informativeness and the aim of faithful representation, *Synthese*, *190*(2), 219–234.

Byrne, D.S. (2001), *Understanding the Urban*, Houndmills: Palgrave.

Byrne, D.S. (2002), *Interpreting Quantitative Data*, London: Sage Publications.

Byrne, D.S. (2005), Complexity, configurations and cases, *Theory, Culture and Society*, *22*(5), 95–111.

Byrne, D.S. and Callaghan, G. (2013), *Complexity Theory and the Social Sciences: The State of the Art* (1st edn), New York: Routledge.

Byrne, D.S. and Ragin, C.C. (2009), *The Sage Handbook of Case-Based Methods*, London: Sage Publications.

Calcott, B. (2008), Assessing the fitness landscape revolution, *Biology and Philosophy*, *23*, 639–657.

Cartier, M. (2004), An agent-based model of innovation emergence in organizations: Renault and Ford through the lens of evolutionism, *Computational and Mathematical Organization Theory*, *10*(2), 147–153.

Castellani, B. and Hafferty, F.W. (2009), *Sociology and Complexity Science*, Berlin: Springer.

Chang, M.-H. and Harrington, J.E. (2000), Centralization vs. decentralization in a multi-unit organization: A computational model of a retail chain as a multi-agent adaptive system, *Management Science*, *46*(11), 1427–1440.

Chettiparamb, A. (2006), Metaphors in complexity theory and planning, *Planning Theory*, *5*(1), 71–91.

Cilliers, P. (1998), *Complexity and Postmodernism: Understanding Complex Systems*, Abingdon: Routledge.

Cilliers, P. (2002), Why we cannot know complex things completely, *Emergence*, *4*(1/2), 77–84.

Cilliers, P. (2005), Complexity, deconstruction and relativism, *Theory, Culture and Society*, *22*(5), 255–267.

Cohen, M.D., March, J.G. and Olsen, J.P. (1972), A garbage can model of organizational choice, *Administrative Science Quarterly*, *17*(1), 1–25.

Conrad, M. and Ebeling, W. (1992), M.V. Volkenstein, evolutionary thinking and the structure of fitness landscapes, *Biosystems*, *27*(3), 125–128.

Cook, S.D.N. and Wagenaar, H. (2012), Navigating the eternally unfolding present toward an epistemology of practice, *American Review of Public Administration*, *42*(1), 3–38.

David, P. (1985), Clio and the economics of QWERTY, *American Economic Review*, *75*(2), 332–337.

Dekkers, R. (2009), Distributed manufacturing as co-evolutionary system, *International Journal of Production Research*, *47*(8), 2031–2054.

Dennett, D.C. (1991), Real patterns, *Journal of Philosophy*, *88*(1), 27–51.

Dervitsiotis, K.N. (2004), Navigating in turbulent environmental conditions for sustainable business excellence, *Total Quality Management and Business Excellence*, *15*(5–6), 807–827.

Dervitsiotis, K.N. (2007), On becoming adaptive: The new imperative for survival and success in the 21st century, *Total Quality Management and Business Excellence*, *18*(1–2), 21–38.

Dietz, F.J. (2000), *Meststoffenverliezen en economische politiek: Over de bepaling van het maatschappelijk aanvaardbare niveau van meststoffenverliezen uit de Nederlandse landbouw*, Bussum: Coutinho.

Dobzhansky, T. (1982), *Genetics and the Origin of Species*, New York: Columbia University Press.

Dooley, K.J., Corman, S.R., McPhee, R.D. and Kuhn, T. (2003), Modeling high-resolution broadband discourse in complex adaptive systems, *Nonlinear Dynamics, Psychology, and Life Sciences*, *7*(1), 61–85.

Dopfer, K. (2005), *The Evolutionary Foundations of Economics*, Cambridge: Cambridge University Press.

Dover, G.A. (1993), On the edge, *Nature*, *365*, 704–706.

Easton, G. (2010), Critical realism in case study research, *Industrial Marketing Management*, *39*(1), 118–128.

Ehrlich, P.R. and Raven, P.H. (1964), Butterflies and plants: A study in coevolution, *Evolution*, *18*(4), 586–608.

Elder-Vass, D. (2005), Emergence and the realist account of cause, *Journal of Critical Realism*, *4*(2), 315–338.

Eldredge, N. and Gould, S.J. (1972), Punctuated equilibria: An alternative to phyletic gradualism, in T.J.M. Schopf (ed.), *Models in paleobiology* (pp. 82–115), San Francisco: Freeman, Cooper & Co.

Elster, J. (1976), A note on hysteresis in the social sciences, *Synthese*, *33*(1), 371–391.

Elster, J. (2007), *Explaining Social Behavior: More Nuts and Bolts for the Social Sciences*, Cambridge: Cambridge University Press.

Epstein, J.M. (2008, 31 October), Why model?, Lecture, retrieved 31 August 2016 from http://jasss.soc.surrey.ac.uk/11/4/12.html.

Ethiraj, S.K. and Levinthal, D. (2004), Modularity and innovation in complex systems, *Management Science*, *50*(2), 159–173.

Fellman, P.V. (2010), The complexity of terrorist networks, *International Journal of Networking and Virtual Organisations*, *8*(1–2), 4–14.

Fellman, P.V. (2011), Using complex adaptive systems tools to solve intractable problems in intelligence analysis and counter-IED operations, 1595–1600.

Fischer, F. (1998), Beyond empiricism: Policy inquiry in postpositivist tradition, *Policy Studies Journal, 26*(1), 129–146.

Fischer, F. (2003), *Reframing Public Policy: Discursive Politics and Deliberative Practices*, Oxford: Oxford University Press.

Fischer, F. and Forester, J. (1993), *The Argumentative Turn in Policy Analysis and Planning*, Durham, NC: Duke University Press.

Fischer, R.A. (1923), On the dominance ratio, in *Proceedings of the Royal Society of Edinburgh* (Vol. 42, pp. 321–341), Edinburgh: Royal Society of Edinburgh.

Foster, J. and Holzl, W. (2004), *Applied Evolutionary Economics and Complex Systems*, Cheltenham, UK and Northampton, MA, USA: Edward Elgar Publishing.

Fox, R.F. (1993), Review of Stuart Kauffman, *The Origins of Order: Self-Organization and Selection in Evolution, Biophysical Journal, 65*(6), 2698–2699.

Freeman, L.C. (1978), Centrality in social networks: Conceptual clarification, *Social Networks, 1*(3), 215–239.

Frenken, K. (2006a), A fitness landscape approach to technological complexity, modularity, and vertical disintegration, *Structural Change and Economic Dynamics, 17*, 288–305.

Frenken, K. (2006b), *Innovation, Evolution and Complexity Theory*, Cheltenham, UK and Northampton, MA, USA: Edward Elgar Publishing.

Garrido, N. (2004), The desirable organizational structure for evolutionary firms in static landscapes, *Metroeconomica, 55*(2–3), 318–331.

Gavrilets, S. (1997), Hybrid zones with Dobzhansky-type epistatic selection, *Evolution, 51*(4), 1027–1035.

Gavrilets, S. (2003), Evolution and speciation in a hyperspace: The roles of neutrality, selection, mutation and random drift, in J. Crutchfield and P. Schuster (eds), *Towards a Comprehensive Dynamics of Evolution: Exploring the Interplay of Selection, Neutrality, Accident, and Function* (pp. 135–162), New York: Oxford University Press.

Gavrilets, S. (2004), *Fitness Landscapes and the Origin of Species* (MPB-41), Princeton, NJ: Princeton University Press.

Gavrilets, S. (2010), High-dimensional fitness landscapes and speciation, in M. Pigliucci and G.B. Müller (eds), *Evolution: The Extended Synthesis* (pp. 45–80), Cambridge, MA: MIT Press.

Geels, F.W. (2002), Technological transitions as evolutionary reconfiguration processes: A multi-level perspective and a case-study, *Research Policy, 31*(8–9), 1257–1274.

Gerrits, L.M. (2008), *The Gentle Art of Coevolution: A Complexity Theory Perspective on Decision Making over Estuaries in Germany,*

Belgium and the Netherlands, Rotterdam: Erasmus University Rotterdam.

Gerrits, L.M. (2011), A coevolutionary revision of decision making processes: An analysis of port extensions in Germany, Belgium and the Netherlands, *Public Administration Quarterly*, *35*(3), 309–339.

Gerrits, L.M. (2012), *Punching Clouds: An Introduction to the Complexity of Public Decision-Making*, Litchfield, AZ: Emergent Publishing.

Gerrits, L.M. and Marks, P.K. (2008), Complex bounded rationality in dyke construction, *Land Use Policy*, *25*(3), 330–337.

Gerrits, L.M. and Marks, P.K. (2014a), How fitness landscapes help further the social and behavioral sciences, *Emergence: Complexity and Organization*, *16*(3), 1–17.

Gerrits, L.M. and Marks, P.K. (2014b), Vastgeklonken aan de Fyra: Een pad-afhankelijkheidsanalyse van de onvermijdelijke keuze voor de falende flitstrein, *Bestuurskunde*, *23*(1), 55–64.

Gerrits, L.M. and Marks, P.K. (2015), The evolution of Wright's adaptive field to contemporary interpretations and uses of fitness landscapes in the social sciences, *Biology and Philosophy*, *30*(4), 459–479.

Gerrits, L.M. and Verweij, S. (2013), Critical realism as a metaframework for understanding the relationships between complexity and qualitative comparative analysis, *Journal of Critical Realism*, *12*(2), 166–182.

Gerrits, L.M., Marks, P.K. and Boehme, M. (2015a), Entwicklung und Scheitern des niederländischen Hochgeschwindigkeitsprojekts 'Fyra', *Eisenbahn-Revue International*, *7*, 340–342.

Gerrits, L.M., Marks, P.K. and Boehme, M. (2015b), The development and failure of the Dutch 'Fyra' high-speed project, *Railway Update*, *9*, 146–148.

Geyer, R. and Pickering, S. (2011), Applying the tools of complexity to the international realm: From fitness landscapes to complexity cascades, *Cambridge Review of International Affairs*, *24*(1), 5–26.

Geyer, R. and Rihani, S. (2010), *Complexity and Public Policy: A New Approach to Twenty-First Century Politics, Policy and Society*, Abingdon: Routledge.

Ghemawat, P. and Levinthal, D. (2008), Choice interactions and business strategy, *Management Science*, *54*(9), 1638–1651.

Ghiselin, M. (2009), Darwin and the evolutionary foundations of society, *Journal of Economic Behavior and Organization*, *71*(1), 4–9.

Girard, M. and Stark, D. (2003), Heterarchies of value in Manhattan-based new media firms, *Theory, Culture and Society*, *20*(3), 77–105.

Grafen, A. (2006), Optimization of inclusive fitness, *Journal of Theoretical Biology*, *238*(3), 541–563.

Green, S. (2015), Can biological complexity be reverse engineered?, *Studies in History and Philosophy of Science Part C: Studies in History and Philosophy of Biological and Biomedical Sciences*, *53*, 73–83.

Hamilton, W.D. (1964), The genetical evolution of social behaviour, *Journal of Theoretical Biology*, *7*, 1–16.

Hantzis, M.A. (2013), *Heavy Metal Leaves a Bitter Taste in My Mouth*, Sacramento, CA: Imaginary Publishers.

Haslett, T. and Osborne, C. (2003), Local rules: Emergence on organizational landscapes, *Nonlinear Dynamics, Psychology, and Life Sciences*, *7*(1), 87–98.

Haslett, T., Moss, S., Osborne, C. and Ramm, P. (2000), Local rules and fitness landscapes: A catastrophe model, *Nonlinear Dynamics, Psychology, and Life Sciences*, *4*(1), 67–86.

Hechter, M. and Kanazawa, S. (1997), Sociological rational choice theory, *Annual Review of Sociology*, *23*(1), 191–214.

Henry, A.D., Lubell, M. and McCoy, M. (2011), Belief systems and social capital as drivers of policy network structure: The case of California regional planning, *Journal of Public Administration Research and Theory*, *21*(3), 419–444.

Hodgson, G.M. and Knudsen, T. (2006), Why we need a generalized Darwinism, and why generalized Darwinism is not enough, *Journal of Economic Behavior and Organization*, *61*, 1–19.

Holland, J.H. (1995), *Hidden Order: How Adaptation Builds Complexity*, Jackson, TN: Perseus Books.

Holland, J.H. (2006), Studying complex adaptive systems, *Journal of Systems Science and Complexity*, *19*(1), 1–8.

Hordijk, W. and Kauffman, S.A. (2005), Correlation analysis of coupled fitness landscapes, *Complexity*, *10*(6), 41–49.

Hovhannisian, K. (2004), 'Imperfect' local search strategies on technology landscapes: Satisficing, deliberate experimentation and memory dependence, Department of Computer and Management Sciences, University of Trento, retrieved from http://econwpa.repec.org/eps/comp/papers/0405/0405009.pdf.

Hulst, M. van (2008), *Town Hall Tales: Culture as Storytelling in Local Government*, Delft: Eburon.

John, P. (1999), Ideas and interests; agendas and implementation: An evolutionary explanation of policy change in British local government finance, *British Journal of Politics and International Relations*, *1*(1), 39–62.

Kakizaki, I. (2014), *Trams, Buses and Rails: The History of Urban Transport in Bangkok 1886–2010*, Chiang Mai: Silkworm Books.

Kallis, G. and Norgaard, R.B. (2010), Coevolutionary ecological economics, *Ecological Economics*, *69*(4), 690–699.

Kaplan, J. (2008), The end of the adaptive landscape metaphor?, *Biology and Philosophy*, *23*(5), 625–638.

Kauffman, S.A. (1993), *The Origins of Order*, Oxford: Oxford University Press.

Kauffman, S.A. (1995), Escaping the Red Queen effect, *McKinsey Quarterly*, *1*, 118–130.

Kauffman, S.A. and Johnsen, S. (1991), Coevolution to the edge of chaos: Coupled fitness landscapes, poised states, and coevolutionary avalanches, *Journal of Theoretical Biology*, *149*(4), 467–505.

Kauffman, S.A. and Levin, S. (1987), Towards a general theory of adaptive walks on rugged landscapes, *Journal of Theoretical Biology*, *128*(1), 11–45.

Kauffman, S. and Macready, W. (1995), Technological evolution and adaptive organizations: Ideas from biology may find applications in economics, *Complexity*, *1*(2), 26–43.

Kauffman, S.A. and Weinberger, E.D. (1989), The *NK* model of rugged fitness landscapes and its application to maturation of the immune response, *Journal of Theoretical Biology*, *141*, 211–245.

Kiblinger, W.P. (2007), C. S. Peirce and Stuart Kauffman: Evolution and subjectivity, *Zygon*, *42*(1), 193–202.

Kimura, M. (1968), Evolutionary rate at the molecular level, *Nature*, *217*, 624–626.

Kindt, D., Cholewinski, M., Kumai, W., Lewis, P. and Taylor, M. (1999), Complexity and the language classroom, *Academia: Literature and Language*, *67*(3), 235–258.

Klein, D. (2015), *Social Interaction: A Formal Exploration*, Tilburg: University of Tilburg.

Klijn, E.-H. (2008), Complexity theory and public administration: What's new?, *Public Management Review*, *10*(3), 299–317.

Klijn, E.-H. and Koppenjan, J. (2015), *Governance Networks in the Public Sector*, Abingdon: Routledge.

Knott, J.H., Miller, G.J. and Verkuilen, J. (2003), Adaptive incrementalism and complexity: Experiments with two-person cooperative signaling games, *Journal of Public Administration Research and Theory*, *13*(3), 341–365.

Knudsen, T. and Stieglitz, N. (2007), Taming the antagonistic forces of exploration and exploitation, retrieved from http://papers.ssrn.com/sol3/papers.cfm?abstract_id=996798.

Knuuttila, T. (2011), Modelling and representing: An artefactual approach to model-based representation, *Studies in History and Philosophy of Science*, *42*(2), 262–271.

Koppenjan, J.F.M. and Klijn, E.-H. (2004), *Managing Uncertainties in Networks: A Network Approach to Problem Solving and Decision Making*, New York: Routledge.

Kurzweil, E. (1980), *The Age of Structuralism: Levi Strauss to Foucault*, New York: Columbia University Press.

Lakatos, I. (1976), Falsification and the methodology of scientific research programmes, in S. Harding (ed.), *Can Theories Be Refuted? Essays on the Duhem–Quine Thesis* (pp. 205–259), The Netherlands: Springer.

Lakoff, G. and Johnson, M. (2003), *Metaphors We Live By*, Chicago: Chicago University Press.

Langton, C.G. (1986), Studying artificial life with cellular automata, *Physica D: Nonlinear Phenomena*, *22*(1), 120–149.

Lansing, J.S. (2003), Complex adaptive systems, *Annual Review of Anthropology*, *32*(1), 183–204.

Lansing, J.S. and Kremer, J.N. (1993), Emergent properties of Balinese water temple networks: Coadaptation on a rugged fitness landscape, *American Anthropologist*, *95*(1), 97–114.

Lasswell, H.D. (1936), *Politics: Who Gets What, When, How*, New York: Whittlesey House, McGraw-Hill Book Company.

Lavertu, S. and Moynihan, D.P. (2013), Agency political ideology and reform implementation: Performance management in the Bush administration, *Journal of Public Administration Research and Theory*, *23*(3), 521–549.

Lee, D. and Van den Steen, E. (2010), Managing know-how, *Management Science*, *56*(2), 270–285.

Levinthal, D.A. (1997), Adaptation on rugged landscapes, *Management Science*, *43*(7), 934–950.

Lewis, D. (2008), *Convention: A Philosophical Study*, Hoboken, NJ: John Wiley & Sons.

Lievaart, P.J. (2014), *De kogel door de Kuip? Een onderzoek naar de agendadynamiek omtrent de toekomst van Stadion Feijenoord*, Rotterdam: Erasmus Universiteit Rotterdam.

Llewelyn, J.E. and Lewis, D.K. (1970), Convention: A philosophical study, *Philosophical Quarterly*, *20*(80), 286.

Losch, A. (2009), On the origins of critical realism, *Theology and Science*, *7*(1), 85–106.

Lovejoy, W.S. and Sinha, A. (2010), Efficient structures for innovative social networks, *Management Science*, *56*(7), 1127–1145.

Luhmann, N. (1977), Differentiation of society, *Canadian Journal of Sociology/Cahiers Canadiens de Sociologie*, *2*(1), 29–53.

Luhmann, N. (1984), *Soziale Systeme: Grundriß einer allgemeinen Theorie*, Frankfurt am Main: Suhrkamp.

Luhmann, N. (1995), *Social Systems*, Stanford, CA: Stanford University Press.

Mackie, J.L. (1980), *The Cement of the Universe: A Study of Causation*, Oxford: Oxford University Press.

March, J.G. (1994), *A Primer on Decision Making: How Decisions Happen*, New York: Free Press.

Marks, P. (2002), *Association between Games: A Theoretical and Empirical Investigation of a New Explanatory Model in Game Theory*, Amsterdam: Thela Thesis.

Masel, J. (2011), Genetic drift, *Current Biology*, *21*(20), R837–R838.

Maturana, H.R. and Varela, F.J. (1980), *Autopoiesis and Cognition: The Realization of the Living* (1st edn), Dordrecht: D. Reidel Publishing Company.

Maynard Smith, J. (1982), *Evolution and the Theory of Games* (1st edn), Cambridge: Cambridge University Press.

Maynard Smith, J. and Price, G.R. (1973), The logic of animal conflict, *Nature*, *246*(5427), 15–18.

Mayr, E. (1963), *Animal Species and Evolution*, Cambridge, MA: Harvard University Press.

McCandlish, D. (2011), Visualizing fitness landscapes, *Evolution*, *65*(6), 1544–1558.

Merry, U. (1999), Organizational strategy on different landscapes: A new science approach, *Systemic Practice and Action Research*, *12*(3), 257–278.

Mitchell, S. (2009), Complexity and explanation in the social sciences, in C. Mantzavinos (ed.), *Philosophy of the Social Sciences: Philosophical Theory and Scientific Practice* (pp. 130–145), Cambridge: Cambridge University Press.

Morcol, G. (2012), *A Complexity Theory for Public Policy*, New York: Routledge.

Moreno-Penaranda, R. and Kallis, G. (2010), A coevolutionary understanding of agroenvironmental change: A case-study of a rural community in Brazil, *Ecological Economics*, *69*(4), 770–778.

Mouzelis, N. (1995), *Sociological Theory: What Went Wrong? Diagnosis and Remedies*, Abingdon: Routledge.

Nelson, R.R. (2006), Evolutionary social science and universal Darwinism, *Journal of Evolutionary Economics*, *16*(5), 491–510.

Nelson, R.R. and Winter, S.G. (1982), *An Evolutionary Theory of Economic Change*, Cambridge, MA: Harvard University Press.

Neumann, J. von and Morgenstern, O. (1953), *Theory of Games and Economic Behavior*, Princeton, NJ: Princeton University Press.

Newton, T. (2003), Crossing the great divide: Time, nature and the social, *Sociology*, *37*(3), 433–457.

Newton, T. (2008), Truly embodied sociology: Marrying the social and the biological?, *Sociological Review*, *51*(1), 20–42.

Norgaard, R.B. (1984), Coevolutionary development potential, *Land Economics*, *60*(2), 160–173.

Norgaard, R.B. (1994), *Development Betrayed: The End of Progress and a Coevolutionary Revisioning of the Future*, Abingdon: Routledge.

Norgaard, R.B. (1995), Beyond materialism: A coevolutionary reinterpretation, *Review of Social Economy*, *53*(4), 475–486.

Odum, E. (2004), *Fundamentals of Ecology* (5th edn), Belmont, CA: Cengage Learning.

Parsons, T. (1991), *The Social System* (new edn), Abingdon: Routledge.

Parsons, W. (1995), *Public Policy*, Aldershot, UK and Brookfield, VT, USA: Edward Elgar Publishing.

Pascale, R.T. (1999), Surfing the edge of chaos, retrieved from http://sloan review.mit.edu/article/surfing-the-edge-of-chaos/.

Peschard, I. (2011), Making sense of modeling: Beyond representation, *European Journal for Philosophy of Science*, *1*(3), 335–352.

Petkov, S. (2014), The fitness landscape metaphor: Dead but not gone, *Philosophia Scientiæ*, *19*(1), 1–16.

Petkov, S. (2015), Explanatory unification and conceptualization, *Synthese*, *192*(11), 3695–3717.

Peyton-Young, H. (1998), *Individual Strategy and Social Structure: An Evolutionary Theory of Institutions*, Princeton, NJ: Princeton University Press.

Pierson, P. (2000), Increasing returns, path dependence and the study of politics, *American Political Science Review*, *94*(2), 251–267.

Plutynski, A. (2008), The rise and fall of the adaptive landscape?, *Biology and Philosophy*, *23*, 605–623.

Poole, M.S., Van de Ven, A.H., Dooley, K. and Holmes, M.E. (2000), *Organizational Change and Innovation Processes: Theory and Methods for Research*, Oxford: Oxford University Press.

Popper, K.R. (2002), *The Logic of Scientific Discovery*, Abingdon: Routledge.

PREGO (Projekt Raum- und Regionalentwicklung Gotthard) (2006), 'San Gottardo: Das Herz der Alpen im Zentrum Europas: Bericht der Kantonsregierungen Uri, Wallis, Tessin und Graubünden an den Bundesrat', Policy document.

Prigogine, I. (1997), *The End of Certainty* (1st edn), New York: Free Press.

Proulx, S.R., Promislow, D.E.L. and Phillips, P.C. (2005), Network thinking in ecology and evolution, *Trends in Ecology and Evolution*, *20*(6), 345–353.

Provine, W. (1986), *Sewall Wright and Evolutionary Biology*, Chicago: University of Chicago Press.

Ragin, C.C. (2000), *Fuzzy-Set Social Science* (1st edn), Chicago: University of Chicago Press.

Ragin, C.C. (2014), *The Comparative Method: Moving Beyond Qualitative and Quantitative Strategies* (rev. edn), Berkeley: University of California Press.

Ragin, C.C. and Amoroso, L.M. (2010), *Constructing Social Research: The Unity and Diversity of Method* (2nd edn), Los Angeles: Sage Publications.

Reed, M. and Harvey, D.L. (1992), The new science and the old: Complexity and realism in the social sciences, *Journal for the Theory of Social Behaviour*, *22*(4), 353–380.

Rein, M. and Schön, D. (1996), Frame-critical policy analysis and frame-reflective policy practice, *Knowledge and Policy*, *9*(1), 85–104.

Reiss, J.O. (2007), Relative fitness, teleology, and the adaptive landscape, *Evolutionary Biology*, *34*(1–2), 4–27.

Rescher, N. (1998), *Complexity: A Philosophical Overview*, New Brunswick, NJ: Transaction Publishers.

Reynolds, C.W. (1987), Flocks, herds, and schools: A distributed behavioral model, *Computer Graphics*, *21*(4), 25–34, retrieved from http://www.cs.toronto.edu/~dt/siggraph97-course/cwr87/.

Rhodes, M.L. (2008), Complexity and emergence in public management, *Public Management Review*, *10*(3), 361–379.

Rhodes, M.L. and Donnelly-Cox, G. (2014), Hybridity and social entrepreneurship in social housing in Ireland, *VOLUNTAS: International Journal of Voluntary and Nonprofit Organizations*, *25*(6), 1630–1647.

Rivkin, J.W. and Siggelkow, N. (2002), Organizational sticking points on *NK* landscapes, *Complexity*, *7*(5), 31–43.

Robertson, B.J. (2004), On moral science: The problematic politics of Stuart Kauffman's order, *Configurations*, *12*(2), 287–312.

Room, G. (2011), *Complexity, Institutions and Public Policy: Agile Decision-Making in a Turbulent World*, Cheltenham, UK and Northampton, MA, USA: Edward Elgar Publishing.

Room, G. (2016), *Agile Actors on Complex Terrains: Transformative Realism and Public Policy*, Abingdon: Routledge.

Rosenhead, J. (1998), Complexity theory and management practice, Department of Operational Research, London School of Economics.

Ruhl, J.B. (1996a), The fitness of law: Using complexity theory to describe the evolution of law and society and its practical meaning for democracy, *Vanderbilt Law Review*, *49*, 1407–1471.

Ruhl, J.B. (1996b), Complexity theory as a paradigm for the dynamical law-and-society system: A wake-up call for legal reductionism and the modern administrative state, *Duke Law Journal*, *45*(5), 849–928.

Ruhl, J.B. (1997), Thinking of environmental law as a complex adaptive system: How to clean up the environment by making a mess of environmental law, *Houston Law Review*, *34*(4), 933–1002.

Ruhl, J.B. (1999), Sustainable development: A five-dimensional algorithm for environmental law, *Stanford Environmental Law Journal*, *18*(1), 31–64.

Ruhl, J.B. and Ruhl, H. (1997), The arrow of the law in modern administrative states: Using complexity theory to reveal the diminishing returns and increasing risks the burgeoning of law poses to society, *UC Davis Law Review*, *30*, 405–482.

Ruse, M. (1990), Are pictures really necessary? The case of Sewall Wright's 'adaptive landscapes', in *Proceedings of the Biennial Meeting of the Philosophy of Science Association* (Vol. 2, pp. 63–77), Chicago: University of Chicago Press.

Ruse, M. (1996), *Monad to Man: The Concept of Progress in Evolutionary Biology*, Cambridge, MA: Harvard University Press.

Samoilenko, S. (2008), Fitness landscapes of complex systems: Insights and implications on managing a conflict environment of organizations, *Emergence: Complexity and Organization*, *10*(4), 38–45.

Sanderson, S. (1990), *Social Evolutionism: A Critical History*, Cambridge, MA: Basil Blackwell.

Savage, M. (2009), Contemporary sociology and the challenge of descriptive assemblage, *European Journal of Social Theory*, *12*(1), 155–174.

Savage, M. and Burrows, R. (2007), The coming crisis of empirical sociology, *Sociology*, *41*(5), 885–899.

Sayer, A. (2000), *Realism and Social Science*, London: Sage Publications.

Schelling, T.C. (1960), *The Strategy of Conflict*, Cambridge, MA: Harvard University Press.

Schelling, T.C. (2006), *Micromotives and Macrobehavior* (rev. edn), New York: W.W. Norton & Company.

Schön, D.A. and Rein, M. (1995), *Frame Reflection: Toward the Resolution of Intractable Policy Controversies*, New York: Basic Books.

Schueler, J. (2008*)*, *Materialising Identity: The Co-construction of the Gotthard Railway and Swiss National Identity*, Eindhoven: Eindhoven University of Technology.

Shotter, J. (1992), Is Bhaskar's critical realism only a theoretical realism?, *History of the Human Sciences*, *5*(3), 157–173.

Siggelkow, N. and Levinthal, D. (2003), Temporarily divide to conquer, *Organization Science*, *14*(6), 650–669.

Simon, H.A. (1969), *The Sciences of the Artificial*, Cambridge, MA: MIT Press.

Sinha, K.K. and Van de Ven, A.H. (2005), Designing work within and between organizations, *Organization Science*, *16*(4), 389–408.

Spekkink, W.A.H. (2015), *Industrial Symbiosis as a Social Process: Developing Theory and Methods for Longitudinal Investigation of Social Dynamics in the Emergence and Development of Industrial Symbiosis*, Rotterdam: Erasmus University Rotterdam.

Spekkink, W.A.H. and Boons, F.A.A. (2015), The emergence of collaborations, *Journal of Public Administration Research and Theory*, *26*(4), 613–630.

Sugden, R. (1986), *The Economics of Rights, Co-operation and Welfare*, London: Palgrave Macmillan UK.

Swoyer, C. (1991), Structural representation and surrogative reasoning, *Synthese*, *87*(3), 449–508.

Teisman, G.R. (2000), Models for research into decision-making processes: On phases, streams and decision-making rounds, *Public Administration*, *78*(4), 937–956.

Tichy, N.M., Tushman, M.L. and Fombrun, C. (1979), Social network analysis for organizations, *Academy of Management Review*, *4*(4), 507–519.

Tollefson, D. (2002), Collective intentionality and the social sciences, *Philosophy of the Social Sciences*, *32*(1), 25–50.

Uprichard, E. (2012), Being stuck in (live) time: The sticky sociological imagination, *Sociological Review*, *60*(S1), 124–138.

Uprichard, E. (2013), Describing description (and keeping causality): The case of academic articles on food and eating, *Sociology*, *47*(2), 368–382.

Uprichard, E. and Byrne, D. (2006), Representing complex places: A narrative approach, *Environment and Planning A*, *38*(4), 665–676.

Urry, J. (2008), Climate change, travel and complex futures, *British Journal of Sociology*, *59*(2), 261–279.

Valente, T.W. and Foreman, R.K. (1979), Integration and radiality: Measuring the extent of an individual's connectedness and reachability in a network, *Social Networks*, *20*, 89–105.

Varela, F.J. (1981), Autonomy and autopoiesis, in G. Roth and H. Schwegler (eds), *Self-Organizing Systems: An Interdisciplinary Approach* (pp. 14–24), Frankfurt am Main: Campus Verlag.

Vayda, A.P. (1983), Progressive contextualization: Methods for research in human ecology, *Human Ecology*, *11*(3), 265–281.

Vega-Redondo, F. (ed.) (1996), *Evolution, Games, and Economic Behaviour*, Oxford: Oxford University Press.

Venkatraman, N. (1989), The concept of fit in strategy research: Toward verbal and statistical correspondence, *Academy of Management Review*, *14*(3), 423–444.

Vishnoi, N.K. (2013), Making evolution rigorous: The error threshold, in *Proceedings of the 4th Conference on Innovations in Theoretical Computer Science* (pp. 59–60), New York: ACM.

Wagenaar, H. (2007), Governance, complexity, and democratic participation: How citizens and public officials harness the complexities of neighborhood decline, *American Review of Public Administration*, *37*(1), 17–50.

Watkins, J. (1989), The methodology of scientific research programmes: A retrospect, in K. Gavroglu, Y. Goudaroulis and P. Nicolacopoulos (eds), *Imre Lakatos and Theories of Scientific Change* (pp. 3–13), Berlin: Springer.

Weber, B.H. (1998), Origins of order in dynamical models: A review of Stuart A. Kauffman, *The Origins of Order: Self-Organization and Selection in Evolution*, *Biology and Philosophy*, *13*(1), 133–144.

Weber, B.H. and Depew, D.J. (1996), Natural selection and self-organization, *Biology and Philosophy*, *11*(1), 33–65.

West, S.A., El Mouden, C. and Gardner, A. (2011), Sixteen common misconceptions about the evolution of cooperation in humans, *Evolution and Human Behavior*, *32*(4), 231–262.

Westhoff, F.H., Yarbrough, B.V. and Yarbrough, R.M. (1996), Complexity, organization, and Stuart Kauffman's *The Origins of Order*, *Journal of Economic Behavior and Organization*, *29*(1), 1–25.

Whitt, R.S. (2009), Adaptive policymaking: Evolving and applying emergent solutions for U.S. communications policy, *Federal Communications Law Journal*, *61*(3), 483–589.

Williams, M. (2009), Social objects, causality and contingent realism, *Journal for the Theory of Social Behaviour*, *39*(1), 1–18.

Winch, P. (2008), *The Idea of a Social Science and Its Relation to Philosophy* (Routledge Classics), Abingdon: Routledge.

Woody, A. (2004), More telltale signs: What attention to representation reveals about scientific explanation, *Philosophy of Science*, *71*, 780–793.

Wright, S. (1931), Evolution in Mendelian populations, *Genetics*, *16*, 97–159.

Wright, S. (1932), The roles of mutation, inbreeding, crossbreeding and selection in evolution, in *Proceedings of the Sixth International Congress of Genetics* (pp. 356–366).

Wright, S. (1968), *Evolution and the Genetics of Populations: A Treatise* (Vols 1–4), Chicago: University of Chicago Press.

Wright, S. (1978), The relation of livestock breeding to theories of evolution, *Journal of Animal Science*, *46*(5), 1192–1200.

Wuisman, J. (2005), The logic of scientific discovery in critical realist social scientific research, *Journal of Critical Realism*, *4*(2), 366–394.

Index